Anonymous

Polaris Investigation

Anonymous

Polaris Investigation

ISBN/EAN: 9783337325435

Printed in Europe, USA, Canada, Australia, Japan

Cover: Foto ©berggeist007 / pixelio.de

More available books at **www.hansebooks.com**

POLARIS INVESTIGATION.

WASHINGTON, D. C., *October* 11, 1873.

Examination conducted on board United States steamer Tallapoosa at the navy-yard, Washington, D. C., of the party under Captain Buddington, from the North Polar Expedition in the United States steamer Polaris, said party having been rescued by the whaler Ravenscraig, carried to Dundee, Scotland, returned to New York City, and brought from that port to Washington by the Tallapoosa.

The names of the rescued party are as follows:
Captain S. O. Buddington.
First Mate H. C. Chester.
Second Mate William Morton.
Dr. Emil Bessels.
First Engineer Emil Schumann.
Second Engineer A. A. Odell.
Seamen Herman Siemans, Henry Hovey, and Noah Hayes.
Carpenter Nathaniel Coffin.
Fireman Walter Campbell.

At 11.30 a. m. Hon. George M. Robeson, Secretary of the Navy, accompanied by Commodore Reynolds, and Captain Howgate, of the Signal Service, assembled at the navy-yard on board the Tallapoosa, for the purpose of taking the statements of the rescued party.

The first witness who was examined was Captain S. O. BUDDINGTON, who, in reply to interrogatories, testified as follows:

By Secretary ROBESON:

Question. Captain, you are aware that, when the party from the Polaris who were on the ice-floe arrived, we thought it proper to examine them and obtain their full statements with a view to preserving everything, not only that the Government may be informed of what has been done and what has been omitted, but that whatever there was of value to history or science might be secured at once; it seems also proper that we should go on with your party in the same way, so that we may have the statements of everybody freely and fully made from their own recollection of what occurred. We have sent for you first as the commander of the expedition after the death of Captain Hall, and we desire you to give a statement, so far as you can, of everything which seems to have any reference to the subject-matter. What is your name?
Answer. Sidney O. Buddington; I live in Groton, Connecticut; my profession is whaling.
Question. How long have you been engaged in that business?
Answer. Since the summer of 1840.

Question. State on what ships, so far as you can recollect.

Answer. The Julius Cæsar, William C. Nye, of New London; the Minerva Smith, of New Bedford; the Franklin, of New Bedford. I sailed two voyages in the William C. Nye. I sailed two voyages in the McClellan, of New London, and four voyages in the brig Georgiana, of New London, and one in the bark George Henry, of New London; also three in the schooner Franklin, of New London.

Question. You commanded some of those ships?

Answer. Yes, sir.

Question. State fully what positions you occupied on board of them.

Answer. I was mate of the Minerva Smith, the Franklin, and the Mc-Clellan, and second mate and boat-steerer in the William C. Nye. I was before the mast in the Julius Cæsar. I was also mate of the McClellan two voyages. I was master of the brig Georgiana four voyages, and of the George Henry one, the schooner Franklin three, and the bark Odd-Fellow one; the Concordia one.

Question. Had you been much in the northern waters previous to this voyage?

Answer. Since the spring of 1850. On the 7th of March I sailed for that country as mate of the McClellan, of New London.

Question. You cruised in what waters?

Answer. We were in Baffin's Bay and Davis Straits, and several times during the season in sight of Cape York, but couldn't get through.

Question. You have never been higher than Cape York before?

Answer. No, sir.

Question. Had you ever spent a winter north in those waters?

Answer. Ten before this voyage. In Frobisher's Bay and that vicinity—what is named now Cyrus Field's Bay. I spent two winters there, and two in what is called Cornelius Grinnell's Bay, and the rest in Cumberland Gulf. I sailed with the Polaris from Washington when she left here as sailing and ice master; went from here to New York, and from New York to New London; from New London to Saint John's, and from Saint John's to Fiskernaes; next to Holsteinberg, and then to Disco. After leaving Disco we went to Upernavik, and from there to a little place where we just stopped, but did not anchor. Captain Hall went ashore and got a few dogs. We anchored at Tessuisak.

Question. Did anything of interest happen after you left New York to the time you anchored at Tessuisak?

Answer. Nothing, except that there was some little difficulty in Disco; Captain Davenport came aboard and had a talk about it. Captain Hall had a very slight difficulty with me about some of my—well, it was a very careless trick in me, and he gave me a reprimand on leaving St. John's. I apologized about it in the best way I could, and there was nothing more thought about it by either him or myself. After we left St. John's Captain Hall came to me about 2 o'clock one night; we were bound for Holsteinberg; he said that he wanted to go into Fiskernaes; we were then nearly abreast of it, and he wanted me to change my course and go there. I got up, looked at the chart, went on deck, and had the course altered; that was all of any note that happened. Nothing of any great importance happened after we left Tessuisak until we got to the latitude which Captain Hall made, 82° 26'.

I have here my general report of the proceedings of the ship, which was written after we were rescued, and while on board the whaler Ravenscraig. I dated and signed it at Dundee, Scotland, the 27th of September, 1873. It contains a general report of the operations of the ship, and of my proceedings, more especially after Captain Hall left the ship to go on his final sledge journey. It contains substantially all that occurred, and I can supply any further information you may desire by my personal examination.

Question. Describe generally what you did after you left Tessuisak.

Answer. We ran to the westward twenty-five miles in a thick fog among heavy bergs; ran very slowly through the night about twenty-five miles; we shaped the course for Cape York, and nothing of any importance happened; we met with a very little ice. The next day, in the afternoon, we raised Cape York not far from 6 o'clock. The next morning we had some little ice to contend with, but none to speak of until we got up abreast of Cape Parry, (or Perry;) there we had quite a lot, but it was very light, and we got through with it quite easily. We passed Cairn Point; from there we shaped the course for Cape Frazier; as we came up to Cape Frazier the ice made in close to the land. Captain Hall landed there and staid a few minutes with the boat's crew. Mr. Chester, I believe, landed with him. There was quite a strong tide setting to the southward at the time, and we were up to the heavy ice then; that is where we first came into the heavy ice. We passed that, and up Kennedy Channel we had no ice at all, nothing to speak of, or to prevent the ship from going where she chose. We had some fog. The next morning after passing through Kennedy Channel the fog was very thick. We stopped to get observations at 12 o'clock, and got some very good ones. Most of us, and myself among them, got 81° 20′. That was the morning after passing through Kennedy Channel. I should have to refer to dates to give them to you. After 12 o'clock the fog lifted, and we had a clear run, with a strong head-wind, however, up through that open place shown on Meyer's chart, which, in my way of talking, I have called "Hall's Basin." It is the widest part of the channel there between the shores. We had passed Cape Constitution clear of ice, and we passed Cape Lieber clear of ice; a little was sticking to the shore, but not much. We passed Lady Franklin's Bay, which we did not see, as we were rather on the other shore. We passed up to Robeson Channel. The wide part is across from what is now called Polaris Bay to Cape Lieber on the west side, and stretching up to Wrangel Bay on the same side.

Question. You have been accustomed to calling the water between what is known as the Southern Fiord and Franklin Bay, which makes the widest part there, Hall's Basin?

Answer. Yes, sir. The ice at Cape Frazier extended pretty close to the shore, but left room inshore to go by, as we passed. We had no more up the channel. We went on with clear water but a head-wind through this basin up into the straits above, not making any stop, except that one, to get our observations. We passed by the mouth of Newman's Bay and saw it was clear of ice. In the morning, about 4 o'clock, we came to a full stop, and Captain Hall told me to go in on the east side. There was no chance to get westward or any way, except into the east from where we were, and he said we would look for a harbor. I accordingly steered square in across the channel for the east side and ran alongside of one heavy floe about six miles, by Walker's patent log, and in the evening the captain tried to land with Captain Tyson; but the tide was running so fast he could not get ashore without losing his boat, on account of the ice. That night we had it foggy. That was at what is called Repulse Harbor, above Newman's Bay. The next morning he tried to land again. We knocked about that night and took every advantage to get north that we best could, but did not succeed. It was foggy during the night—somewhat misty. Next morning Captain Hall landed again with Mr. Tyson—or rather tried to land at the same spot, but they couldn't get ashore. They came near losing their boat, as was reported to me. It was all I could do to keep the shore there, the heavy floe was going so fast. There was a little water inside, but it was a very strong tide. It was about the full of the moon, I think; I do not recollect now, exactly; but I know there was a very strong tide running to the south. We had the ship under steam, and when Captain Hall came aboard he took a look at it, and I recommended strongly that we should go to Newman's Bay, which would be an open place. That was about eight or ten miles south of us. That was the only place I could see, and I thought it best to go in there, and if the channel cleared we could see it and be open to it, and not run the risk of getting beset in the ice in trying to stay out there. He held a council with the officers— Dr. Bessels and myself, and the others—which I have here, that was written down as it occurred, I believe, word for word. (Paper is now marked " No. 1, B.")

It read as follows:

" Consultation held on board the Polaris in regard to getting further north with the vessel, the vessel being on the east side looking for a harbor. Dr. Bessels, Mr. Meyers, Captain Tyson, Captain Budding- ton, Mr. Morton, and Mr. Chester. Doctor wanted to cross the straits to look for a harbor, as being better for sledge journeys, while the east side was better for navigation, if we could not get further north. Mr. Mor- ton coincided with Dr. Bessels; Mr. Meyers had the same opinion; Mr. Chester to get as far north as possible; Captain Tyson to get into harbor as soon as possible; Captain Buddington to keep on east side as being better for navigation, and certainly better for sledge journeys. It was impossible to get further north on account of the pack. Go along the coast on the east side of the straits southward until a harbor is reached, which could be done in a short time. There had been seen one a few miles to the south of present position of the vessel. It was decided by the commander to cross the straits. In doing so we got beset by the pack and drifted back about fifty miles."

Captain BUDDINGTON, (resuming:) That paper was written down at the time, and it was the same in Captain Hall's journal, which, unfortunately, has been lost. It was left on the ice.

Question. Is this paper in your handwriting ?

Answer. No, sir; it was written by my instructions. It is a record of the consultation and opinions given at the time, written down by my instructions by Captain Hall's clerk, perhaps about a week after it oc- curred. The same thing was written down by Mr. Meyers in Captain Hall's journal. Captain Hall once read it to me from his journal, and I got the clerk to write down a copy of it, which is this copy. There was something said, I believe, on board the ship about it not being a proper way to do business, or something of that kind, and I spoke to Captain Hall about it. This statement is entirely correct, not only according to my recollection at the time when it was written down, but according to my recollection now, and according to my recollection from Captain Hall's journal, and it is according to my recollection of the facts as they occurred.

Question. What happened then ?

Answer. Captain Hall told me to cross the straits, and if we could not succeed we would come back and go into this bay, now called New- man's Bay on the chart, but in doing so we got beset in the heavy ice and drifted away to the southward. We were drifting part of the time against quite a strong southerly wind.

The 4th of September, at night, we got out. The ice slacked up and we had steam on. We had landed some provisions during this time in a heavy floe, but had got them back on board. We were fast between two heavy floes, and we landed all our deck-load, pretty much, and a good deal from below. As the ice began to slack up we took it aboard again, and on the evening of the 4th of September the ice slacked up so that we succeeded in getting inshore and got into the lead of water that made inside from Cape Lupton. There was always water there when there was any anywhere; we had it all summer there, more or less. After the ice first broke we could not get back into Newman's Bay though. The ice swept up the bay to Cape Lupton. Captain Hall went on the ice and looked for himself, and told me there was no use to try to go up any further. We could not get up Robeson Channel on account of the ice, which swept close in to Cape Lupton; it did so all the summer. I tried several times to get up there, and I always came to a stop. Getting back to the ship you would think you could go right by by the looks of the water, but as soon as you would get up, there was a block. Finding we could not get up we went to landing our provisions and stores and made ready to winter there. This was the place which Captain Hall called "Thank God" Harbor. We landed pretty much everything that we wanted to land and made all the room in the ship that was necessary. We put up the observatory on shore. .

During the month of September Mr. Chester and the doctor went on a musk-ox hunt, and returned, I think, on the 26th.

Afterward a party went to the southward, Mr. Meyers among them, a short journey of one day to get some bearings of the other shore. They were sent by Captain Hall. Some of the Esquimaux were out sealing a good deal of the time and general work was going on. Until Captain Hall started north on a sledge journey, things were going on very quietly and appeared to be all right.

I here produce the orders which he left me.

The said orders, comprising eight pages of foolscap, are now marked "No. 2, B," and read as follows:

"UNITED STATES STEAMSHIP POLARIS, C. F. Hall, commanding.
(Official.)

"SIR: I am about to proceed on a sledge journey for the object of determining how far north the land extends on the east side of the strait in which the Polaris is wintering, and also to prospect for a feasible inland route to the northward for next spring's sledging, in my attempt to reach the north pole; this route to be adopted providing the ice of this strait should be found so hummocky that sledging over it should be impracticable, and furthermore to hunt musk-cattle, believing and knowing as I do, from experience, that all the fresh meat for use of a ship's company thus situated—as is that of the Polaris—should be secured before the long arctic night closes upon us.

"You will as soon as possible have the remainder of the stores and provisions that are on shore taken out under the plain by the observatory and there placed with the other stores and provisions in as complete order as possible. You will have each kind by itself as near as may be. You will have the ship's houses (winter awnings) put up as designed. Have the night-watch kept up in accordance with my written instructions of September 23, with merely this change: that the watch is to be continued until the cook commences his morning work. Have every light in the ship extinguished at 9 p. m.; except from this hour a candle-light is to be allowed forward for the use of the watch.

" You will see that no more coal is consumed in any stove of the ship than actually is necessary. I find by the thermometer, placed in the men's quarters forward and both cabins aft, that the temperature of the air is kept far higher than it should be, both for economy in the consumption of coal and for the health of the ship's company, the thermometer through the day and evening ranging from 60 to 70 degrees. Therefore you will require no more coal shall be consumed than is necessary to keep the thermometer, forward and aft, at 50° through the day and evening. A very small fire to be allowed forward, to be kept up from 9 p. m. through the night; but the one aft to be discontinued at 9 p. m. Have the dogs well cared for, feeding them every other day. Look out some good warm place in the ship for the puppies, and have them well nursed.

" Have Mr. Morton get and open one can of pemmican, and deal it out economically to the puppies.

" I have great hopes of securing many musk-cattle on my sledge journey, and then we can spare much of our ship's provisions to the dogs. Should any such calamity be in store for the Polaris—which I pray to God may not be—that a storm from the northward should drive the ice out of Thank God Harbor, and the Polaris with it, during the coming spring-tides, then have steam gotten up as quickly as possible, and lose no time in getting the vessel back again to her former position; but should the Polaris be driven into the moving pack-ice of the straits and become beset, and you should not be able to get her released, then, unfortunately, the vessel and all on board must go to the southward, drifting with the pack, God only knowing where and when you and the ship's company will find means to escape. It might, in this case, be that such a drift movement would occur as in the case of the United States Grinnell expedition of 1851-'52, and of the Fox, under McClintock, 1857-'58; and whenever you should get released, if anywhere between Cape Alexander or Cape York, or between the latter and the arctic circle, you will then make your way to Godhaven, in Disco Island; and if the Polaris remains sea-worthy you will fill her with the coal, stores, and provisions, and the next fall, of 1872, steam back to this place.

" If the vessel should become a wreck, or disabled from the imminent exposure and danger of such an ice-drift as referred to, then all possible use of your best judgment must be brought into play for the preservation of the lives of all belonging to the expedition.

" You will, at your earliest moment of escape, acquaint the Government of the United States with the whole of the circumstances; and should one of those circumstances be the loss of the Polaris, I, and my small party that is absent to accompany me in the proposed sledge journey, would remain here to make discoveries to the north pole, making Thank God Harbor our headquarters, and all the time feel certain that our country would lose no time in sending us aid in carrying out the great object of the expedition.

"Although I feel that it is almost certain that the Polaris is safely lodged in her winter position, yet we know not what a storm may quickly bring forth. A full storm from the south can send the pack of the strait impinging upon the land-pack in the midst of which we are, and in a few minutes cast the Polaris high and dry upon the land.

" During the coming spring-tides let great vigilance be exercised, especially during any gale or storm, at the time of high tides.

"As soon as time will allow, have snow-blocks cut from the drifts under the lee of the hill by the observatory, and sledged over to the Polaris, the same to be placed about her as an embankment.

"You will have plank and boxes so placed under the poop that the dogs cannot get to the raw-hide wheel-ropes.

"The usual routine of the ship that I have established will be gone through with each day during my absence. You will see that this is carried out, including church-service each Sabbath.

"The duties that devolve upon Mr. Morton, by my appointment, are that of paymaster and yeoman. He has full charge, under my direction, of all the accounts, stores, and provisions on board the Polaris, and on shore, belonging to the United States.

"Whatever relates to the consumption and use of said stores and provisions, Mr. Morton has charge of and will be made responsible for the same. I am sure this trust which I have committed to Mr. Morton will be carried out with fidelity, and to the best advantage of the service of the United States Government in this its North Polar Expedition.

"All the fuel, kindling, and coal before being used must pass through the hands of Noah Hayes, who must keep an exact account of the same, which he must vouch to Mr. Morton, or he may render the amount to the chief engineer and the latter to Mr. Morton; no box, barrel, or anything else, furnished with provisions, must be opened by Mr. Morton. So far as these and all other orders I have issued, you will have carried out. You will keep a journal of all proceedings during my absence and transmit the same to me on my return. You will not omit to note such violations of orders that are or may be given, and by whom; nor will you omit to note the conduct of any and all. Hoping that God will protect you and help you in the discharge of the duties that devolve upon you, I bid you adieu and all those of my command, trusting on my return to find all well, and trusting, too, that I shall be able to say that my sledge journey, under the protection and guidance of Heaven, has been a complete success, having made a higher northing and nearer approach to the north pole than any white man before, and that a practicable inland sledge route further north has been found, that many musk-cattle have been seen and captured.

"I have the honor to be your obedient servant,
"C. F. HALL,
"*Commanding United States North Polar Expedition*,
"*Latitude* 81° 38' *North, Longitude* 61° 44' *West, October* 10, 1871.
"To S. O. BUDDINGTON,
"*Sailing and Ice Master*, *United States North Polar Expedition.*"

Captain BUDDINGTON, (continuing:)

The captain went on his journey. He started, I think, about 2 o'clock in the afternoon. He encamped five miles from there. Next morning he sent Hans back after some things, a pair of seal-skin pants and one or two little things which I think he has mentioned. Part of his request was mentioned in that little book. He sent me a letter, too, which I have not got, and I wrote back to him particulars and sent the things. Hans left, I think, about 1 o'clock that day. I sent him a letter that everything was right, and detained Hans a short time to have the pants fixed. He was gone until the 24th of October. I never heard of him any more after Hans left the ship till he returned. In the mean time everything was very quiet, and I was at work according to those instructions. He returned about 2 o'clock on the 24th of October. I saw the sledge coming, and I went ashore to meet him, and met Mr. Chester first and asked him how everything was; they said they were all well, had a good time, &c. When I came to Captain Hall he was very lively. I met him on the shore at the top of the bank. He gave up his sled then and was walking down. I came up to Chester first, who was just ahead of him a short distance. I walked off to the ship with him. He spoke very favorably of the land journey, and thought that there was a good chance to travel some ways north on that side, and spoke of seeing signs of musk-ox; had shot two seals in Newman's Bay, called it an uncommon good harbor, and wished he was in it. He appeared lively. I asked him if there was a good road up this side to get north. He says "Yes; I can go to the pole, I think, on this shore." We were then housed over and banked up with snow on the outside, and a door to go in over the rail. He went in, and I went to my work on the outside of the ship, banking up with what men we had on the outside, putting snow around for winter protection. In about an hour and a half, more or less, after that, he sent out for me. I went in, and he was then in his bunk, and said he felt a little sick coming in out of the cold, and had been vomiting slightly. He told me to have the sleeping-bags dried and everything put in order; that he wanted to go south day after to-morrow and look at this fiord that is now on the chart, and wanted Mr. Tyson to go with him. He said he thought he had a bilious attack, and inquired of me if I didn't think he needed an emetic. I told him "yes." Dr. Bessels stood by, and said it would not do for him to take an emetic. I had given him a great many first and last. Even at home, a few days before we went away, I had given him an emetic. He was very subject to those bilious attacks.

Question. Did you say that he said he was sick from having come in out of the cold?

Answer. That is what he spoke about. He said he felt sick at the stomach and had been vomiting, from coming in out of the air. He said he thought he was bilious and had better take an emetic, and asked me if I didn't think so. Dr. Bessels was present, and he said he didn't think it would do for him to take an emetic. He gave no reason that I know of. If he did, I don't really know what it was. I don't recollect what he said about it, only it wouldn't do for him. It was not far from two hours after he came on board—an hour and a half or so. I should say an hour and a half safely. Mr. Morton was attending to him, and he was getting some clothes for him to shift his clothes when he went to bed. When he came back from going into his state-room to get the clothes he told me that he was abed then. I don't know how long he had been abed when I came aboard. From that time he grew worse. I kept on with the regular work as it had been under his instructions, and he would at times be perfectly rational and then he would give some orders that I attended to, and kept on from that to the 8th of November. He was sometimes out of his head. I think it was two days before he was taken down the last time he was, as I supposed, entirely well. Even the last day he told me, "I shall be in to breakfast with you in the morning, and Mr. Chester and Mr. Morton need not sit up here with me at night. I am as well as I ever was." He dictated some writing that day to his clerk. He read a good deal, and in the afternoon he had quite a nap in his chair. Hannah was with him. That night I went to bed at the usual time. He was sitting up. I slept alone. I saw him sitting there. Shortly after 12 o'clock, I think it was, Chester aroused me up and says, "Captain Hall is dying." I ran up as quick as I could. He was sitting in the berth, with his feet hanging over, his head going one way and the other, and eyes very glassy, and looking like a corpse—frightful to look at. He wanted to know how they would spell "murder." He spelled it several different ways, and kept on for some time. At last he straightened up and looked around and recognized who they were, and looked at the doctor. He says, "Doctor, I know everything that's going on; you can't fool me," and he called for some water. He undertook to swallow the water, but couldn't. He heaved it up. They persuaded him to lie down, and he did so, breathing very hard. When I first went up I asked Mr. Chester what he had been taking. He said the doctor gave him something just as he was going to bed, and he went right to bed and went to breathing in this way, very hard. It appeared to be not exactly a snore, but between hard breathing and a snore all the time. This was along after 12 o'clock, between that and 2. The next morning he came out, and appeared to be as strong as he ever was. As quick as he got up they called me, and I went in. The doctor was in there, and he looked around among us and wanted me particularly to note down what he said, as it would be interesting when he got well.

Question. Who wanted you to do that?

Answer. Captain Hall; he was crazy—out of his head. He stayed up for a few minutes and called for some more water; tried to drink it, and couldn't swallow it. He lay down again and went to sleep; that is, I supposed he was asleep by the way he was breathing. He never got up again, and died that night, somewhere along about 2 o'clock.

Question. Do you know whether he took any medicine that day?

Answer. Nothing but injections, as I understood. I never saw them, but he never took any medicine during the day. I understood the doctor used an injection, as he said, of quinine. He told me so. The doctor, I mean, told me so.

Question. When Captain Hall talked in this way, which you have detailed, was he in his right mind?

Answer. No, sir; not all the time. When he talked about being in to breakfast and dictated some writing, &c., he was in his right mind—perfectly well. That was the night before he was taken down.

Question. I mean after he was taken down.

Answer. No, sir.

Question. He never appeared to be after that?

Answer. No, sir; not after I was called up that night by Mr. Chester, and went up. He never was in his right mind. He only spoke twice after that.

Question. Did he seem to have an idea that people were poisoning him, or murdering him, or something of that kind?

Answer. Yes, sir; always. He insisted upon it.

Question. Did he accuse anybody in particular?

Answer. Yes, sir.

Question. Who?

Answer. Dr. Bessels. At times he thought everybody was at it. But he appeared to spit out his whole venom on him; that is, he appeared to think that the doctor was the proper one.

Question. At times, you say, he seemed to think that everybody was doing it?

Answer. Yes, sir.

Question. Who else did he accuse besides the doctor?

Answer. I believe the cook, one night. He told me that the cook had a gun in his berth and was going to shoot him. The cook slept on the opposite side. I went over and overhauled the berth; there was nothing there, of course. One night I was up in the cabin and Captain Hall got hold of me pretty severely. I called to Mr. Chester and Mr. Tyson to come up. They were down below. Captain Hall knew what I said, and held me by one shoulder, and took hold of the door-knob and held it tight, and they had some difficulty in getting the door open.

Question. What did he take hold of you for?

Answer. I really could not tell. He was just in that way. He said that there was blue around the lamp, and blue gas coming out of my mouth, and everything of that kind. He said also the same to Mr. Tyson. He would feel for his mouth, when he was close by him, and say "What's that coming out of your mouth? It is something blue." One night Mr. Chester or Mr. Morton, I do not recollect which, put a pair of stockings on him—the others were wet; he had stepped in the water. He objected to it very strongly, and said they were poisoned; but they finally persuaded him to have them on. I believe that Hannah came in and put the stockings on. He would not allow anybody else to do it.

Question. Did he take his meals all this time?

Answer. No, sir; he never went to the table, but he used to eat quite freely, so I understand, but I don't know. I was not in a great deal of the time. I generally went when I was called for. I had a good deal to do at the time. Mr. Chester, Mr. Morton, the doctor, and the Esquimaux woman were attending to him. He used to take some wine, I believe, and I saw tamarind-water. He used to drink that, so they told me, and I saw it in the dishes around there.

Question. Who lived in the cabin with him?

Answer. The engineer, Dr. Bessels, Mr. Bryan, Mr. Myer, the cook and steward. Morton, Chester, Tyson, and myself were below with the Esquimaux family and the second engineer.

Question. What was the condition of Captain Hall's mind during his first attack?

Answer. About the same as the last one, only not so bad. He was quite delirious one spell, and he got partially over it, during the first attack, and he sent in for Mr. Chester, Mr. Morton, and Mr. Tyson. He was not exactly right then, but nearly so. He told them then that he wanted to give up the care of the ship to me entirely; that the crown fitted him too tight; that he had enough to attend to his surveys, and he didn't want to be bothered any more with the ship or crew. Mr. Myer lay in his berth, I believe, at the time and heard it. Apparently anybody who was not acquainted with him would have thought that he was entirely rational then. He was not, however. Before that he had been out of his head a good deal. One night he got up really desperate, and Mr. Chester was watching with him. He called me. I went up. He was fairly raving. I tried to get him to bed, but I could not do anything with him. He said we had all joined in with that little German dancing-master to disgrace him, and he was perfectly ready to leave the world. But it did not last but a short time before he went to bed, and he was apparently quiet. I understood him, by that expression, to mean Dr. Bessels. The doctor was there at the time, I think.

Question. Did the doctor say anything?

Answer. No, sir. He tried to pacify him, as the rest of us did, whenever he got in that way. I would not be sure that the doctor was in there. I know Mr. Chester was, and I really think the doctor was. I am nearly certain of it.

Question. What did you think was the matter with him?

Answer. I thought it was a bilious attack that first occasioned it. I thought I had seen him in the same way before, and doctored him accordingly. We were up in those latitudes together twenty-seven months before.

Question. Had you seen him out of his head before?

Answer. No, sir; the way he was taken first is what I refer to.

Question. Did you take all these charges and sayings of his to be the expressions of a man in his right mind, or the expressions of a crazy man?

Answer. Of a crazy man.

Question. Was the effect that they made upon you at the time the effect of a man in delirium?

Answer. Yes, sir.

Question. Was that the impression they produced upon you?

Answer. Yes, sir.

Question. What was the impression which you understood at the time they produced upon the other people?

Answer. I think it must have been the same to all those who heard it.

Question. Did anything occur at that time, which came to your observation in any way, which induced you to believe that anybody was trying to poison him or trying to injure him in any way or shape?

Answer. Well, sir, I don't think there was, only the doctor came to me one night and says, "Captain Hall is quite unwell, and won't take anything." I said, "Can't you get him to take something?" and says I, "Doctor, mix up a dose more than you want him to take, and if he sees me take some of it, he will take it then without any difficulty." The doctor said, "It will not do for you to take the first drop of quinine." That's all the remark I heard. And he said once to me that he thought Captain Hall was a physician, but he knew then he was not? Says I, "How do you know that?" "Well," he says, "he didn't know that quinine could be injected into the system." That's about all I heard. He spoke once saying that he had a very strong constitution, &c.

Question. You were present when Captain Hall died?

Answer. Yes, sir. I was called when he was quite bad, and I stopped there. It was only a very few minutes before he died. I was up the biggest part of the afternoon and evening, when he was lying there. He was entirely insensible from about 9 o'clock in the forenoon until he died, about 2 o'clock the next morning.

Question. Was he breathing very hard?

Answer. Yes, sir. Sort of snorting through the nose. His eyes were very glassy. After he died he looked very natural, indeed. He was buried on the 10th of the same month, in the day-time. We took advantage of what little light there was. He died at about 2 o'clock on the morning of the 8th.

Soon after Captain Hall's death Dr. Bessels drew up this document which I now produce, and which is signed by myself and him, dated at Thank God Harbor, November 13, 1871.

(The document is now marked "No. 3, B," and reads as follows:)

"*Consultation.*

"THANK GOD HARBOR,
"*November* 13, 1871.

"First consultation held between Messrs. S. O. Buddington and E. Bessels. Through the mournful death of our noble commander, we feel compelled to put into effect the orders given us by the Department, viz:

"'Mr. Buddington shall, in case of your death or disability, continue as the sailing and ice master, and control and direct the movements of the vessel; and Dr. Bessels shall, in such case, continue as the chief of the scientific department, directing all sledge journeys and scientific operations. In the possible contingency of their non-concurrence as to the course to be pursued, then Mr. Buddington shall assume the sole charge and command, and return with the expedition to the United States with all possible dispatch.'

"It is our honest intention to honor our dear flag, and to hoist her on the most northern part of the earth, to complete the enterprise upon which the eyes of the whole civilized world are raised, and to do all in our power to reach our proposed goal.

(Signed)   "S. O. BUDDINGTON.
(Signed)   "EMILE BESSELS."

He was buried about a third of a mile south of where the observatory stood. We put up a head-board and fixed the grave up as well as we could. (Producing envelope.) Here is a piece of the willow which came from it. We planted a willow shrub there; it was doing very well when I took that from it, the last time when we were there, the day before we left. We left on the 12th day of August, 1872, about nine months after his death. After his death we went on regularly, and I followed his instructions as near as I could and as long as I could. The scientific men kept on with their observations, and two men from the crew were taking note of the tide. Everything went on regularly until the latter part of the month of November, when the ship broke out and went alongside of the berg, and we made fast to it during the gale. After the gale was over, and left the ice free a little, we got somewhat clear of it a little distance from the berg, and I let her lay. We sawed the ice clear from her some distance from the berg, and the reason I did not take her any further from the berg was that in case we broke out again I wanted it to hold on to. This was a large berg grounded there in from 12 to 13 fathoms of water, the same berg which Captain Hall had called the "Providence Iceberg," and under the lee of which we had made our harbor. A few days after I had sawed her clear, the very last of the month, there was a heavy southwest gale, and the pressure of ice on the outside of the berg drove it afoul of the ship. The ship lay on the northeast side of the berg, between the berg and a little bend of the coast. Thank God Harbor was made by a little bend in the coast-line, with the berg on the outside, the bend of the coast on the north-east side of the berg. The southeast gale pressed the berg in against us, and the tongue of the berg ran under the ship about 40 feet from forward to aft. That pressed her inshore through the ice, and piled it up astern and inside of her, and doubled the ice under. As it broke down it would run under her, and piled up very high. The tongue of this berg when it went under her wrenched her awfully, and it also started the stem. The shoving of her in through that heavy ice wrenched the stem so that you could hear everything crack; every timber, especially forward, appeared to be giving way. It wrenched her very badly; I could not get her clear. I was over on the ice a great deal of the time when it was pressing her in, until it got so that the water made in the gangway, and I got a board through that and sta'd there until the berg stopped coming. Afterward I took a look around the ship, and I saw it was impossible to saw her out. She had got to go astern about 40 feet, and the ice was piled high up there. Really, if the ice had not broken astern of us, I do not know how we should have got away. We had to lay on there during the winter. Every low tide, the full and change of the moon, that inshore ice would lift, and the berg being grounded, gave her a keel of about two feet during the lowest part of the tide, and when it was up she would be on an even keel; that, of course, wrenched and strained her very badly. The stem was broken that night I spoke of; that is, I judged so from the crack; it was an old crack when I saw it; I found it first myself. We remained there in that way during the winter, carrying on the observations and doing whatever was necessary to be done. Herman Siemans and Robert Kruger, I think, were taking tidal observations part of the time. The others were doing whatever there was necessary to do on board the ship; everything appeared to be going on very well.

On the 24st of February, I received this letter and inclosure from Dr. Bessels.

The letter is here marked "No. 4, B," and reads as follows:

"WINTER-QUARTERS,
("Latitude 81° 38' North, Longitude 61° 44' West,) February 21, 1872.

"SIR: As, with the return of the sun, the further operations of the expedition must be begun, and as, in regard to all these, a consultation between us should take place, I forward herewith to you the sketch of a plan by means of which, as I think, we may best fulfill the mission upon which we are sent.

"Very respectfully,
(Signed)   "EMIL BESSELS.
"Captain S. O. BUDDINGTON,
        "United States Steamer Polaris."

The inclosure, entitled sketch of plan of operations, is now marked "No. 5, B," and reads as follows:

"As matters stand now there are two ways of accomplishing the object of the expedition; either by boats and the vessel herself, or, as at first proposed, by sledges. Let us, now, consider both ways and the plan of operations for each that seems to offer the most advantages.

"The setting-out of a boat-party will, of course, depend entirely upon the area of open water and the improbability of new ice being formed that would interfere with its navigation. Perhaps, the party could start during the last of March or in the beginning of April—that is to be seen—if the vessel does not break out before that time, which may occur at any time, as our anchorage does not give us much protection.

"If the journey toward the north should be made by means of a boat, considerable time must elapse before it can be safely begun, and the question arises how to employ that time to the best advantage.

"As the object of the expedition is a geographical one, and as geography consists not merely in laying down a coast-line, as many may think, but requires much more than that, a sledge-party should be formed, provisioned for twenty days, to penetrate into the interior of the country, to discover if it consists of an ice-plateau, as is supposed by some, but which does not seem probable, or, in a word, to investigate its configuration. This would also give an opportunity for answering some important questions contained in the instructions.

"Another party could, at the same time, go to Cape Constitution, to determine astronomically the position of Morton's furthest point, which, in regard to longitude, ought to be verified. Besides that, these points of the coast-line should be connected with the survey of our anchorage.

"Regarding the matter of verifying positions, it will also be very desirable to send a party to Grinnell Land, the coast-line of which, although changed a good deal by Dr. Hayes, does not seem to be correct, and ought to be resurveyed. Besides that the party could, perhaps, find out if the land contained any glaciers, as Dr. Hayes stated.

"There is no doubt that it would be considered as a very valuable geographical discovery to determine how far Grinnell Land extends from east to west, which might be done by ascending some of the high mountains near its coast. It must be confessed that this party would be subject to many difficulties and much risk, even if open water did not impede their progress, because the ice is rough and hummocky, and liable at any moment to go adrift.

("As matters stand since the day before yesterday, it would be impossible to cross the strait. February 21, 1872.)

"It is not impossible that the ice in the southern part of the straits will be better for traveling purposes, so that the Cape Constitution party might cross with comparatively little difficulty, but if you take into consideration how much trouble it cost Dr. Hayes, who crossed the strait twice, how it enervated his party, it seems better to give up this plan, especially because next summer there would be very likely a more convenient way of reaching Grinnell Land.

"As it has been concerted, the Polaris will leave at her anchorage a depot of provisions and a boat. Should the vessel be compelled to leave her anchorage before the sledge parties return, then the party arriving first at Polaris Bay should wait for the other, and upon its arrival proceed to Newman's Bay, (the only harbor we know of toward the north,) in the most expeditious manner. By all means it would be a good plan if the vessel breaks out before the return of the sledge parties to leave also a boat with a patent log and provisions at Newman's Bay, because the boat left at Polaris Bay would be used to carry the united sledge parties, and there should be another to fall back upon, in case of accident.

"If the vessel should drift south during the absence of the parties, then documents of the further route they intend to take will be found a few feet to the west of the present site of the observatory. The spot may be known by the iron bar which now holds the pendulum-case.

"Let us return, after this digression, to consider a plan for the operations of a boat party toward the north. One of the smaller boats should be taken, with as many provisions as possible, the necessary instruments, and small stores. The party should follow up the eastern side of the strait, surveying the land and making such investigations in hydrography, in regard to currents, sea atmosphere, and soundings, as may be made without too much delay.

"As near each full degree of latitude as possible the party will build a cairn, and deposit a record of its proceedings, in order that the vessel, if necessary, may know where to search for it.

"Should we, notwithstanding the favorable prospect we now have, be compelled to use sledges on the journey toward the north, then we should start as soon as possible, by all means by the middle of March, because it is not probable that then the temperature will be much lower than it is now, although we might have more gales.

"It cannot be denied that it is a great advantage to use dogs for draught, provided sufficient game can be procured on the way for their food, but as we are compelled to travel over a poor country and make large distances the dogs will prove hindrances rather than help. We must, then, as the English expeditions have done, almost exclusively use men for draught. Two dog-sledges should be taken, loaded with four small sleds, the provisions belonging to them, and besides provisions for the whole party for thirty days. Should the two sledges meet with many difficulties in advancing, which will very likely be the case, then they will establish, at places they may find favorable, small depots of provisions for their return, stay as long as possible with their small sleds, and return when circumstances require it. Then the small sleds will be loaded with the undiminished provisions, and each man drag his own sled, a total weight of two hundred pounds.

"By no means can the small sleds expect to return by the same way over the ice, because at that time it will be broken up, and the vessel herself under way for a high latitude.

"As has been mentioned in the case of the boat party, the sledge party will also build cairns and deposit records of their proceedings.

"Having arrived on their return at a place from which they are unable to travel any further south, they will keep up a continued watch, and signalize, by flags and smoke, while the vessel fires a gun several times a day.

"Now, a few remarks upon the operations of the vessel. It would undoubtedly be best to use as little as possible of our coal, and to proceed north by sail. If it is possible for the vessel to advance along the coast of Grinnell Land it would be profitable to do so, on account of the running survey that could be made, as there certainly will be some one on on board who can conduct a work of this kind.

"The determination of the local attraction of the compass before the vessel starts should not be neglected as heretofore, because without this an able survey cannot be made.

"It should be considered as a matter of the highest importance to take deep-sea soundings, or soundings in general, whenever practicable ; for, except those made by John Ross in 1818, there are but a few taken by Inglefield, and two by Kane. If the time will not allow of more, one sounding a day would be valuable, and should be taken.

"If the water is not very deep one of the smaller sledges should be used to procure a larger number of specimens than can be obtained by the apparatus of Brooks.

(Signed)   "EMIL BESSELS.
" *Winter Quarters, Latitude* 81° 38′ *North ; Longitude* 61° 44′ *West.*
"FEBRUARY 10, 1872."

On the 29th of February I sent to Dr. Bessels the following reply, which was written by Mr. Meyer from my dictation. This is a copy in Mr. Meyer's handwriting. The same was here marked "No. 6, B," and reads as follows :

"THANK GOD HARBOR,
" *February* 29, 1872.

"SIR : I have carefully examined the contents of your communication dated Thank God Harbor, February 10, 1872, and your suggestions as to an early trip to Cape Constitution and the inland meet with my entire approval. Anything to the furtherance of science which can be done before the starting of the final expedition to the north, in pursuit of the principal object of this expedition, I would decidedly advise you to undertake, and you may be assured that all possible aid on my part shall be given to you and your undertaking. The expedition to the north will, within all probability, proceed by the aid of boats ; and it is my decided intention in such a case to take command of the boat party. To come to any conclusion as yet in regard to the details of this boat journey and the proceedings of the ship appears to be useless, inasmuch as circumstances will generally govern our actions.

"Very respectfully, yours,
(Signed)   "S. O. BUDDINGTON,
                        "Commanding United States Steamer Polaris.
"To Dr. EMIL BESSELS,
        "Chief of the Scientific Party of the North Polar Expedition."

What I meant to refer to in saying I would take charge of the boat party myself, was to put such men into the party as I thought proper.

Secretary ROBESON. You say here you meant to take command yourself?

Answer. Yes, sir. What I meant by that was to put Mr. Chester and Mr. Tyson into those boats to take charge of them, as I had nobody else. The instructions which I gave to them are in the journal.

Question. What you meant, then, was that you would take command of their organization and direct their proceedings?

Answer. Yes, sir. The orders I gave that boat party are pasted in the journal. After this the doctor made a sledge journey to the south, and got back not far from the 8th April. Mr. Bryan went with him, with Joe and Hans. A sled broke down, and Mr. Bryan and Joe came back and repaired their sledge, and left the doctor and Hans down there, and then they went back again; when the doctor came back he was nearly snow-blind; they were gone a fortnight. I had but very little to say about the sledge-journeys, as we had come to the understanding that the doctor was to have charge of those and the scientific operations. But I kept everything prepared, and even had the small sleds and big sleds made for him. I got everything as near ready as I could, and kept the dogs in good order and condition. I didn't see much prospect of the sledges going north, and I proposed a sledge journey to Dr. Bessels for Joe and Bryan; they were very anxious to go; they were really the best ones, as I considered. I proposed it to Dr. Bessels, and he asked if I had said anything to Mr. Bryan about it; I really equivocated a little; I didn't like to say no or yes, for I had spoken to Bryan about it. But, however, I got over it in some way without telling a very bad falsehood, and then he told me that he would speak to Mr. Bryan about it. It went along, I think, for a day or two. I asked Mr. Bryan every once in awhile, and sometimes twice a day, if the doctor had said anything to him, and he told me he hadn't. One evening Mr. Bryan spoke to him and told him he was all ready to start in the morning, and wanted to know if he had every preparation made, &c. The doctor rather resented it, and, I thought, made use of some language that was not called for. I heard it from below, and went up into the cabin, and the doctor said he wanted two sleds immediately. I said, "All right, doctor; when do you want them?" He said the first fair day. I said, "All right," and then I went and told Joe and Mr. Bryan that it was no use for me to try and do any more; that the doctor had charge of the sledge journeys, and wanted one or two sledges, and I gave up then undertaking to do anything about it. I took the boats from the shore alongside of the ship the 1st day of April. Dr. Bessels sent off Mr. Meyer and Mr. Tyson, as I understood, on a musk-ox hunt. But still he was going in the boats when they were ready. The Esquimaux went too, and Meyer went along under Bessels's instructions, and was gone some time. I believe his report, rendered to me when he returned, is somewhere in the papers. They were gone quite a while, and, as I understood, reached the latitude of 82° 9'. When they came back I asked Captain Tyson why he didn't go further, as they had killed twelve musk-ox during the cruise among them, and had plenty of dog-food and whatever they chose to have from the ship. He said he was out of fuel. I asked Meyer why he didn't go further, and he said that he went as far as he wanted to ; that he didn't choose to go any further.

On the 1st of April we took the boats from the shore alongside of the ship on the ice, and Mr. Chester went to work on his to fix it to suit himself. Captain Tyson, with the carpenter, went to work on his. There were two four-oared boats. They got them in perfect order to suit themselves for the expedition, and I supposed and understood from the others that we should have to make such a journey, or undertake it, if at all, with boats. The sledding was then wearing away very fast. The snow was going from the land, and it would be bare in a very short time if they didn't take advantage of the spring. They fixed them up and were ready. I told them to have them ready by the 1st of May, and they would have been ready. They had every convenience that could be got up, even boxes for the chronometers, a sounding-line of 2,500 fathoms on each boat, reeled up on the stern and all rigged for that purpose. Somewhere after the 1st of June we sledded those boats up to Cape Lupton. I can't exactly recollect the dates. Mr. Chester started first, and got a short distance, when he unfortunately lost his boat, with nearly everything in it. The ice ran over him in the tide-way. It was no fault of his. He was doing as he thought best, and as well as any man could do under the circumstances. He came back to the ship, and Captain Tyson got an opening after that, and got up as far as Newman's Bay before Mr. Chester overtook him. Mr. Chester got back to the ship and rigged the canvas-boat and started again. We could not spare the other two boats in case they might be lost. After Mr. Chester left that time I heard no more from them for some time.

The ice broke on the outside of the berg, and I found that my saw which I had made here could not work. I cut one of the saws over that day, rigged derricks, and went to sawing three or four days and nights as long as we could stand it, and sawed the ship out clear of the berg and got under way immediately. I supposed the boats had got some ways north, and I wanted to overtake them if possible.

We had a very heavy gale of wind. The ice came up ; there was no going any further, and I could not have used steam ; if I had I would not have done much better, the gale was so heavy ; I carried sail very hard and got her up some distance ; the pack was forming and the gale still continuing ; I ran back under the berg waiting for a better chance. I made three faithful trials and found I could not get any further north. I only used steam once. After the first trial up, two of the men came back from Newman's Bay after some provisions—Herman Siemans and Robert Kruger. Then I sent Hans away with a note to Mr. Chester, advising him to come aboard, as, if there was no chance to get north, we might save some fuel, and I thought then we could get north under the circumstances as quickly as the boats could. But he could not get aboard with the boats, and he sent back word to me to send him back the men. Finally, the doctor came back with Hans and staid aboard after that for some reason or another. I don't know why he came back. I wrote to Mr. Chester, as I say, to that effect, that I thought he had better come aboard and we would try to pump the ship, and, if there was an opening to get north, we would go. He sent word back by Hans that he wished I would send one of his men back at any rate, that it would be some time before he could get down, the way the ice was. I sent them back with what provisions they could carry, and sent him

another letter, telling him how the ship was situated, and if he saw
chance to go; I should recommend him to go by all means, or something to that effect. After that I heard no more from them, but they
came scattering back. Tyson came first after landing his boat on
shore. I left it to their option to do after that as they thought proper.
They went away the 8th of June, I think. Mr. Chester left the ship at
that time, and he got back somewhere about the 22d of July. They
were gone from a month to six weeks. He and Herman Siemans were
the last to come back. They staid to get their boat ashore. She was
on the ice when the others left her.

Question. What did you do then after they came back?

Answer. I really could do nothing; there was no chance to get the
ship away. Mr. Chester came back about the 22d of July, and I never
got a chance to get away from there. We had good lookouts from the
top of the hill every day, and somebody was up there with the glasses.
Hans was my main dependence for that. He was very faithful and
trustworthy, reporting the ice, and even took pains to draw a chart
of where the water and ice was while he was there. I found that I
could not stay another winter, with what coal we had, and a leaky
ship, and I spoke to Dr. Bessels to give it to me in writing. It was his
opinion that we had better come home, as it was mine, and I wanted to
have it in writing, but he put it off, and I finally didn't say much more
to him, or nothing more. We sawed her out clear of the berg, and
started on the 27th June to go north. I had to work pretty hard sawing. My hands were so blistered that I could not hold on to the ropes.
Every man was at work on board except the second engineer, who was
attending to the donkey-pump. When I was out there I had no seamen aboard, with the exception of Morton, and one other to steer.
That was the cook; the others didn't know anything about it. I did
all I could to get the ship north, but failed. When we were trying to
get north the first time under sail, we had the wind right down northeast. The southerly winds, as soon as it breezed up at all, would set
the ice into Polaris Bay, so that it was impossible to get out of it at
all. The northeast winds opened the ice most where we were, but they
drew right down this channel above us. Where we lay they drew off
the land, and as you came to Cape Lupton the ice was tight. Laying
in Polaris Bay you would see the ice outside of you, and anybody
would naturally suppose you could go up, but the ice swept by the
point so that the boats could not get down. It was so close to the
shore from Newman's Bay I could not get up. I thought I saw we
should never do much more there by staying there another winter in
taking observations, and one year was about enough, as I considered, to
take what observations we could there. I didn't see that I could get
any further north.

After the whole of the boat-crews got aboard, which was about the
22d July, Mr. Chester and Herman Siemans being the last, we did
nothing but watch the ice for a chance to get out. Before they
came, it took me all the time to keep her afloat. She was jammed
ashore four or five times. After they came, I had men enough to man
the pumps with proper reliefs, and then I waited my chance to get out.
On the 12th of August we started. We steamed down to that little
island marked on the chart, near the west shore, and there we
stopped for a short time, and drifted through between the two
islands that you see on the chart, and after getting through a little ways, we got another lead and worked down to where we were
finally beset. My orders to the officer on deck were to work in to the

westward with every lead they could find; but they were too favorable to the eastward, and we could not find any lead to the westward; every lead appeared to favor a little more to the eastward. We either had to stop in our progress or take those leads; but everybody was anxious to get along, and, finally, we came to a block about 12 o'clock of the 15th of August. I think it was in the latitude of 80° 2'. We made several trials to get her in-shore from there, but without success. Every time I tried, it cost me some 4 or 5 tons of coal, and I saw the thing was impossible to get her away from there. She finally froze in. We drifted gradually through the month of September and averaged a mile or two a day. After we got down toward the straits further, we took a faster drift, and, finally, drifted to the southward of Cape Alexander. We were somewhat south of that when the heavy gale took us which parted us. We were anchored to the floe with the best hawsers we had, fore and aft. We were beset at the floe, I think, the night of the 15th of August. We made a few trials and got a lead west from there, but nothing to speak of. From the night of the 15th August until the night of the 15th October we were fast to the floe. We had put up a tent on the ice, and framed it up by mortising into the ice, and got it covered with the awning. We landed some bread and things we could spare, in case we all went adrift from the ship, and had everything else on deck ready; whips were all on the yard; the boats both on that side; and we made every preparation we could, really, to land on the floe. We had a sail drawn under the ship's bows, tight under her keel—a piece of this awning—and picked all the old rope into oakum and shoved down through there, and took ashes from the galley and put it on top. But in these nips she got; she tore all clear of the bows and left her leaking badly, and it took the stem, from where it was cracked in the first winter, clear out entirely. All that was gone. If she had not been an uncommonly good ship she would have sunk right there; but the lining inside kept her afloat. She got some very heavy pressures that night. She was very strong in the bow, but was not exactly the right shape. Where the crack was she was two feet through as far aft as the crack run. She was almost as strong as a ship could be made and stood as much, I suppose, as any ship could stand; she stood more than most of them would. Few ships would have stood the pressure she did that night without going down immediately.

We had drifted down below Cape Alexander that time, and I thought the ship was going clear and would come down Baffin's Bay, as the party did on the ice, and then get out in the spring all right and everything safe. I felt encouraged when we got to the south of Cape Alexander, and I am certain we should have done it if we had not got that heavy gale of wind from the southwest. At that time I had drifted the whole way down Smith's Sound, coming through the straits in that tideway and getting down in the north water among the bergs there. I knew we had the water to contend with, more or less. We generally do have every year. Then we took this heavy gale of wind. It had been blowing about twenty-four hours before the 15th. About 6 o'clock on the evening of the 15th of October it was blowing and snowing very thick. They reported to me in the cabin that the ice had cracked on her starboard side—the side opposite the big piece that we had our house on. I went out and found that it had opened then about two feet from her side and was setting off very fast. As the wind was very near aft, I had an extra warp put out aft and hauled the one forward tight. Soon afterward she got a heavy pressure that came up on her starboard side and heeled her over to port so that the rail was nearly onto the ice.

Then it was reported to me by Mr. Schuman, the engineer, and Captain Tyson, that she was making water very fast, and I ordered them to get the provisions and other things onto the ice immediately, which they did, and carried them back as fast as they could. I hurried them up to get them back from the edge of the ice. There was one point astern of us that had some on it and I was afraid it would crack off, as it did. Some of the men were on the ice and some on the ship. They were on the ice a greater part of the time. I noticed when the ship was getting a heavy pressure of ice grinding her there would be more on the ice than aboard, and when she slacked up there would be more aboard. Mr. Chester worked very hard, and so did everybody else. Not far from 10 o'clock I looked and saw one of the aft anchors had jammed out, and the cleets aft that I had used for the purpose of making the hawsers fast to were torn off, and both hawsers were made fast to the mainmast. I saw one of the ice-hooks was out on the floe, and I sent a man to cut another hole to put it in and another one onto the house to slack up that hawser that was fast to that hook. While I was looking at him cutting the hole I saw the ice had cracked on the point where he was putting in the hook. I turned again and looked forward and saw that the whole stem, where the warps were fast, was gone. That is to say, I saw that the ice to which the stem was fastened had broken off so that she was loose at the stem from the floe. I cast my eye forward and saw that the warp was gone, and before I could say anything we were all gone—that is, the whole ship was loose. A·boat lay across the crack and one man said, "What shall we do with it?" I said, "Haul her out on the high hummocks;" and that was all I could say to them. We were out in the darkness in a minute. I tried to get a light up, but I could not keep a light in the ship at all, it was blowing so hard. We flew away there for quite awhile. I could see the ice to leeward going as fast as we were. Our propeller-wheel was full of ice, frozen solid, jammed up. The machinery was all frozen, too. We had no water in the big boiler and no fire under the small one. The fire was built under, but the water was not hot. The engineer came and said that we would have to do something; that the water was getting into the fires as fast as possible. We rigged the pumps on deck by turning warm water into them out of the boiler. They started easily. We had been pumping with the bilge-pump and had kept free with that for a number of days. We got the deck-pumps rigged and all hands got to pumping. We hove blubber and pine doors and everything of that kind in to get steam up. We kept the big pumps going. We could not get the water off the decks. The scuppers were all frozen, and the water was above our knees at the pumps before it would go out of the ports. We could not keep the scuppers clear on account of the snow and ice on deck. We managed to keep the water from the fires finally and got the steam up and the steam-pumps going, and then we kept her free during the night. We were drifting then to the northeast, until in the morning, when it became nearly calm and the gale nearly died away. When we parted from the ice we must have been about abreast of Littleton Island. As near as I could tell, I thought I recognized a berg that we went very close to, fast to this floe that lay there all winter. Littleton Island is above Cumberland Island, about sixty miles, I think. Before the snow set in I could just see Northumberland Island— just before it was going to snow. That was two days before we separated. I think it was on the 13th. I could just see Northumberland Island in the morning; then we drove back before the wind with that gale until we were nearly abreast of Littleton Island, I should judge.

After we went adrift we drifted right into the northeast as long as the breeze lasted. In the morning I unbent the foresail before daylight. It was a square foresail. I set the men all to making bags before daylight to get out the coal from below, so that we might use it if we had to go on the ice. We had no other canvas. As it became light I saw we were north of Littleton Island about three miles, and about three miles from the shore, or not far from that. There was no lead when daylight came. I could not spare coal to get steam on in the big boiler, and I thought at one time it was best to take some of the rigging off the mast, there being more than we wanted, and get the fire started in that way and then burn some coal. I had just decided, knowing I had to work quickly as we were slowly drifting off shore, when the wind came to the northeast and the ice began to slack right in the direction we wanted to go and no other. I got sail on and got a little steam on the small boiler. We would get a three-minute turn before the steam would run out, and I worried her inshore. We got her within about twice the ship's length of the shore, I should say, or perhaps a little less. We came to the shore-ice as she took the bottom. The ice was frozen to the shore. It was the top of high water and the full of the moon. At low water she was hard and fast and fell over on her side as far as she could go. Next day we took a look at her bows and found it was impossible to do anything with her under the circumstances; and we saw it was too far along then to repair her, if we had had the means, and could have got her to a proper place even.

We took our spars and sails ashore and went to work at the house, landed all the provisions and everything that was in her pretty much, and made her fast and solid where she lay. We had great help from the Esquimaux in doing this with their sleds and dogs. Next morning after we got there, two came. They worked all day. They came from Etah, about six miles off, on the mainland. I really don't know how they come to discover us first. The first I saw was early in the morning; they were hallooing, and they came off and proved to be some of Dr. Kane's acquaintances. I set them right to work and they worked faithfully, and went back again and promised to bring the rest. I told them to bring all the men there was there. According to what I had heard of them I thought they would not come. I paid them very well, and the next morning seven came with six sleds; they worked faithfully as long as we stopped there and as long as we wanted them sledding the things ashore. We got ashore on the shore-ice; it was grounded where the ship took the bottom. That point was the only place along there—I believe about the only place where we could have got anything on the beach. When the gale died away the next morning, after we broke adrift, we used our sails. We only used the steam for pumping, and what little we could spare to help her around a point or through a streak of light ice or something of that kind. We had no water in the big boiler at all and had no steam on except from the small one, and we got some help on that to help her along. Sometimes we would cut through a point of ice where it was light. It took us from the time we got started until 4 o'clock in the afternoon to get in there, about three miles. I think we started about 9 o'clock.

Question. Could you recognize the floe at all from which you had broken adrift?

Answer. No, sir; I had the best lookouts at the mast-head, Mr. Chester and Henry Hobby; they could not see anything even with the best glasses in the ship. I had them up at daylight looking for the men; they could not see them. They thought they saw some provisions on the ice, but they could not tell. We kept a regular lookout for them but never saw them. When we started to come down we thought we were about three miles north of Littleton's Island and about three miles from the shore. We came down under sail with what little assistance we could get from the steam with the small boiler and went in about two miles north of Littleton's Island, at a point known as Lifeboat Cove, of Dr. Kane's. Our vessel grounded about twenty rods from the beach, and just there we reached the ice and were able to get ashore by the shore-ice.

Question. At the time you turned in toward shore there you had these lookouts at the mast-head?

Answer. Yes, sir; Mr. Chester was aloft nearly the whole time from the time we started in, after we got started first, until we got ashore.

Question. You did not see these men nor any signs of them?

Answer. No, sir.

Question. How was the atmosphere?

Answer. Quite clear, that day.

Question. No wind?

Answer. It breezed up in the afternoon quite strong; northeast.

Secretary ROBESON. About what time did you make the shore?

Answer. Not far from 4 o'clock; it was just getting night in that latitude. I judge that was the hour; I didn't look, but I think I have heard them say it was about 4 o'clock. I had considerable other things to think of at that time instead of looking.

Question. If you had seen the men could you have got to them in the ship?

Answer. Never, sir. It was all she could do to get ashore. If we had been out another night I don't know what the result would have been, though we had a moon and perhaps could have worked somewhere. We might have got to a heavy floe. The ice we were in then was not fit to land anything on. We had either to get to a heavy floe and get onto it, or else get to the shore, or the ship would sink. We couldn't pump her to keep her afloat; she would have sunk in a very short time.

Question. Had any of the Esquimaux seen the men on the ice at all?

Answer. No, sir. I inquired of every party that came. I thought they might get ashore to the south of us, and then, by the aid of the Esquimaux, I should have found out and got them. I made particular inquiry of every party which I found out came from Cape York, or near there and up to where we were, and I came to the conclusion the party was not on that shore at all, and that possibly they might be picked up on the other shore, as McClintock and De Haven had drifted down there, and swept close into Cape Walsingham. Whalers generally have done so. When Chester thought he saw the provisions on the ice, it was to the southward of us. Whether he was mistaken or not I cannot say. I thought there would be dogs, and he would see them running about. I asked him if he saw any, and he said he did not. There must have been some dogs there on this piece where the provisions were; possibly there might not have been any, but the last I saw nearly all the dogs were on the ice.

Question. How high was that ice above the level of the sea, that they were on?

Answer. I should think the highest part may have been fifty feet.

Question. Did Chester see this piece with the provisions on it in the morning?

Answer. Yes, sir.

Question. Were there any high hummocks or peaks of ice between you and where this was?

Answer. Yes, sir; to the south of us appeared to be all ice. Down for a little ways it would appear to be completely tight; as near as he could tell, it seemed a firm pack. There were a great many bergs grounded, and some hummocky ice. Bergs were very plentiful.

Question. How high are these hummocks?

Answer. Some would be from twenty to fifty or a hundred feet high. There were some very high places on the floe the men were on, and some very low. Where the house was, it was between two heavy ridges.

Question. Was there any refraction in the atmosphere that morning?

Answer. I think not, sir; it appeared to be a very clear morning.

Question. How far were you from shore, according to your estimate, when you broke loose from the ice?

Answer. Really I couldn't tell, but I shouldn't think we were but a short distance from Littleton's Island, by the bergs. When it came morning I could not see any bergs any distance from Littleton's Island off-shore. Chester thought he recognized a berg we were close to that night, and one or two others, and I think I did while we were drifting fast to this floe. We passed very close to it. Then a few minutes after we broke adrift we were by a large berg, very close; so I should say we were quite close into Littleton's Island when we broke adrift. Hans, once before we went adrift, thought he saw the land close to us, but I think it was a berg.

Question. How do you account for not seeing the people on the ice?

Answer. Really I cannot say, sir. The lookouts, I am sure, did the best they could to see them, and how it was I do not know. I shall never be able to tell, unless they were behind some hummock or some berg. If they saw us, however, we ought surely to have seen th m.

Question. You think if you had seen them you could not have reached them?

Answer. No, sir; we would not have been able to have got to them. I supposed, with the boats and everything of that kind, that they would have been able to have got to us better than we could get to them. I saw one of the men in New York, and he said that the pack-ice was so bad that they could not get away, though they tried to. But I really could not get that ship where they could not get with a whale-boat. If I should have got steam on and tried to steam to them I should have torn her all to pieces, and then perhaps never got more than half way to them. I asked the second engineer that day how much coal it would take to get steam under the big boiler. He said nearly all. We found that there was from five to six tons when we got it out of the bunker.

When I ran the ship on shore two miles north of Littleton Island at Lifeboat Cove, she took the bottom just about the time she took the shore-ice, so that we had direct communication with the shore. We commenced immediately to hoist out the coal below that remained there and getting the stores ashore. We took down the sails and spars to make a house, and Mr. Chester commenced on the house with some assistance, and I commenced getting the coal, provisions, and other articles ashore and made the ship fast and solid where she lay to the grounded ice on the shore with heavy hawsers and cables. After that we drove ahead as fast as we could and finally got everything arranged for the winter. Got a galley-stove and cook-stove on deck and took the cook-stove ashore. This was the only low place along there. It so happened that we got in there at the only place we could have got in. There we lived very comfortably through the winter.

It was twenty-four hours or thirty-six hours before we saw the natives. Only two made their appearance at first. I set them to work
that day, the second after we reached shore. They worked all day, and
at night when they went home I paid them for their services with
knives and files, and I gave them a couple of sword-bayonets. Next
morning they brought in seven natives and six dog-sleds and they all
went to work then. We got everything of every character out of the
ship. We took all the coal, all the provisions, all the implements, and
all the records that I knew of, on shore. The records were in a box.
There were a few of them that were not taken ashore at that time, but
they were taken ashore during the winter. The ice held firm all winter.
We were about twenty rods from the shore. I think we were about
four days at it, from the time we began, in getting the things on shore.
Meanwhile the house had been built of spars and bulk-heads of the
rooms below torn out of the ship; we used the ceilings in the rooms
below. The spars were laid on and covered with sails, double. We were
fourteen, all told. We then made arrangements for the scientific operations, and some two or three men went over to Etah on a hunting excursion to try to get some deer; that is an Esquimaux settlement, about six
miles from where we were, and I believe it is the most northern one on
the coast. There are a few houses only there, but nobody inhabited
them while we were there. We had flour, bread, Indian meal, some oat-
meal, some potatoes, canned meats, salted meats, and dried apples. We
had a good supply, but the only thing we had more than we needed
was dried apples and potatoes. A great deal of provisions had been
lost overboard at the breaking of the ice, I suppose, though I don't
know how much. It was all put on the ice; I suppose some one put it
there, but I did not see it. I never saw much of any that was lost. It
was afloat on the ice the last I saw. There was a great deal put overboard from the ship at night, and when the ship broke away some of it
broke apart. The part that was aft broke away.

I have got a list of very near the whole of what was put on the ice here;
(presenting a list;) the list was entitled as follows: "The following is a
list of provisions and articles put upon the ice before the ship broke
away, as near as I can come to it." This list is written by Joseph Mauch,
the man who acted as clerk, and written out at my dictation soon after
we got on the shore. I had a pretty good run of all the provisions, and
I took down a list of them, as near as I could tell by what was left.—
(The list above referred to was marked by the secretary "No. 7, B.")
There was none of the pemmican left; it was all put on the ice.

The next day after we grounded, Mr. Chester, myself, and several
others, and, I believe, the doctor, examined the ship; also Mr. Bryan, and
I do not know but they all did; but both Mr. Chester and I made an
official examination to see the condition she was in. Her stem was gone
from the 6-foot mark down about as low as we could see, and the wood
ends torn off and a good many of them broke. The 6-foot mark is six
feet from the bottom of the keel, up. It is the 6-foot draught mark;
from that point the whole front part of the ship was gone. The planks
were sprung out; the hole was not through, if it had been she would have
sunk immediately, of course, but the ship was lined inside; that and
the timbers kept her afloat. The space was clear down as far as we could
see, about three feet; that was entirely gone and showed what must be
below from what was above. The wood ends of the plank had sprung
out from the stem. The stem below was broken and gone; we lost that
part of our stem the night of the nip, the 15th of October; the stem
had been broken before. We had sails on the bow, and oakum was shoved

in between the sails; that was all carried away by the storm. She got a heavy pressure from the ice, and one of the seamen told me at the time that the stem must be gone. Some considerable damage was done there before the 15th of October.

Part of it was gone the last we saw of it before that, which was when we were getting the sail on the bows; perhaps from the 1st of October to the latter part of September; I don't recollect exactly. She had not got pressure between that time and the night we parted on the ice. I suppose the stem went that night. The ship did not get nipped after she had broken loose from the ice that night. She drifted and came up to light ice after that, and the wind died away shortly afterward. I could not see any means of keeping her afloat under the circumstances. I think if we had had coal, and any kind of clear weather, we might have managed to pump her out and have got her afloat and taken her somewhere else, perhaps, where we might have fixed her; but under the circumstances we could not get anywhere else, and we could not fix her there. We had one small boiler, with an engine, aboard ship, which we used for pumping, which would burn about as much coal as a small stove. We could not keep her free without using steam. The night we went adrift we were pumping as hard as we could, and were just able to keep her up long enough to keep the water out of the fire until we got steam up under the small boiler.

I did not hold any formal survey over the ship to see if she were sea-worthy, under the circumstances, as the ice laid around her so that we could not; she was frozen clear round so that there was no chance to examine her at all. After the first day she was really in a condition where we could not do anything; we could not have done anything more with her, under the circumstances; we had only from five to six tons of coal; we could not any more than have got steam up with that; we went into winter-quarters; we did not have any consultation previous to that as to what to do; we did all that was left to do. It was everybody's opinion to do then what we did do. The coal that we took ashore we used as fuel. We had one small stove and another cook-stove besides. The scientific operations were carried on during the winter. Hourly observations were taken nearly all the time with the thermometer and barometer and as to the velocity of the wind.

I think it was four nights I staid aboard after we got the ship ashore, while we were taking the things out. The ship's crew staid ashore one night before I went there. The ship at low water was very much on one side, and we could not stay aboard very well, and went ashore at once. We got some fresh meat during the winter from the natives. I believe we never got any hares or rabbits in the fall at all. Two or three parties went out, but were unsuccessful. They got several foxes during the winter around the house. We got no more musk-oxen; there were none down there at all. We got one reindeer in the spring, and several rabbits, large white hares. We got no fish. We got some large seals from the natives, and several parts of seals. The natives visited us during the winter from Etah and below. I think they came from the lower settlements toward Cape York. I think all the men, at any rate, that were there that winter, and the biggest part of the women, came to see us. I have got a list of something like a hundred who were there. That list is in the journal. They did not speak English. There was a family from across on the west side came from down where the whalers had been. I could converse with them understandingly, and they interpreted to us. I could not converse with the Cape York natives at all. This family from the west side staid with

us all the winter. The family consisted of a man, wife, and two children. The man went a good deal to hunt. I don't know whether he got
anything much. They did not know Joe. They came from farther north
than he did. We staid there until the 3d day of June. We had intended to leave the first, but that was Sunday. Monday it blew too hard.
Tuesday we left. I informed the whole party that I wanted to be ready to
leave the 1st of June, and we made that arrangement. We commenced
building boats in April. The coldest month we had there was February; that is, I think the thermometer ranged the lowest that month.
It began to moderate about April. There were some mild days in
March, but very few. The boats were built of the ceilings of the cabin.
By the ceilings I mean the inside walls, the stanchions, &c. Mr. Chester
conducted the boat-building; the carpenter and John Booth helped him,
and some of the rest—Mr. Odell and others. The boats were 25 feet
long, and 5 feet beam amidships, and flat-bottomed. They had no keel.
We used masts and sails. They sailed very well with fair wind.

Nothing notable happened during the winter, other than I have detailed, that I recollect. We kept a log that contains a general summary
of everything that happened. We left Lifeboat Cove on the 3d day
of June. Our idea for leaving so early was to get to the whalers. I
never expected to get any farther than Cape York to meet the whalers.
I knew they always came that way and had for years. We came pretty
near being too late for them as it was. Seven had gone down before we
got there, and there were only three others. It is the custom of whalers
to follow up the fast ice through Melville Bay by Cape York on the east
side; and sometimes they have to go to the north of the Carey Islands,
and then cross over the west side to the mouth of Lancaster Sound. We
expected to catch them as they went up on the east side. We had a very
passable sort of a time, and got down there very slowly but surely. We
stopped at Hackluyt Island, which is off Northumberland Island; from
there we got on to Northumberland Island and stopped one or two nights.
After leaving that island we came across to Cape Parry and came along
the east shore, across the sound to Cape Parry, and worked along that
shore. The next move we made we got down to Cape Walsingham.
We stopped at Dalrymple Island, adjoining Walsingham. From there
we got across to Conical Rock. That is still farther down on the east
side toward Cape York—I suppose about fifteen or twenty miles north
from it. We stopped there two nights and not far from two days before
we got started again. Then we were on the ice until the 23d of June,
and then we raised this ship about twelve miles off. We were then
about twenty-five miles south of Cape York. There was a good deal of
ice between us and Cape York. The fast ice was in here. We did not try
to get in. We made a straight course along the fast ice from Cape
York. We did not put in because I was aware that the ships kept the
fast ice alongside of them, knowing that the ships always kept the fast
ice, and coming up till they got so they could cross over to the westward. We raised this ship about twelve miles off, and soon after seeing
her we saw men coming toward us on the ice from her. They had seen
us first. We had hauled up our boats on the ice, coming to a stop, and
went into encampment there, waiting for an opening, when we raised
this ship, and soon after saw these men coming. She was twelve miles
off, we considered it. The men took what they could, and we took
what we could, and we left the boats and went to her. We called it
twelve miles; it was a pretty tough way to go. We were very kindly
received aboard the Ravenscraig and taken across. She got relieved
on the 4th of July from where she was, and got out into the sailing

water, and crossed over to Lancaster Sound with both steam and sail. The Ravenscraig was a bark of about 400 tons, with steam-power, commanded by Captain William Allen, from Dundee. The ship hailed from Kirkcaldie, but sailed from Dundee. When we got over to Lancaster Sound, we fell in with the Arctic. After the first seven men had gone aboard the Arctic the rest remained at that time aboard the Ravenscraig. A short time afterward we fell in with the Intrepid, another bark with steam-power, Captain Senter, also from Dundee. Three of our men, Mr. Bryan, John Booth, and Mauch, went on board of her. The Ravenscraig had poor accommodations for so many men, and we were afraid the provisions would not hold out, and that is the reason they had to be provided for otherwise.

On the 20th of August following, the Arctic had completed her cruise. She had a very successful one and was ready to go home; and these men who were left on board of the Ravenscraig were taken on board of her. We sailed on the 1st of September, and reached Dundee, I think, on the 18th, but I am not positive. We left the Intrepid off Cape Clyde, I think, or very near there. She was whaling; she lacked very little at the time of being ready to sail. 1 expected as soon as she could get another whale she would be going home. There were two other ships beside the Intrepid there that would bring the men home. The Eric only lacked one whale of being ready, and the Intrepid about the same. I never spoke to the men on the Intrepid after I went on board the the Arctic. She was in sight all the time; she was not in any dangerous position at all; all the danger of the voyage was supposed to be entirely over. The Clyde is nearly opposite Discoe on the map, on the western shore. The whaler would remain north of Cape Walsingham until the middle of October; then they will probably be carried down to Dundee by the Intrepid. When we parted with them they were all in good health. Mr. Booth was fireman. Mauch acted as clerk. Captain Hall took him for that purpose; and he did all the writing I had to do afterward. We arrived at Dundee on the 18th; then we were taken care of by the United States consul there and supplied with everything we needed. We went from Dundee to Liverpool, and from Liverpool sailed in the City of Antwerp.

I have a journal—here it is; there are several places there to which I wish to call your attention. (The journal is produced, entitled, "The journal of S. O. Buddington; written by J. B. Mauch, from the death of the commander, Captain C. F. Hall, to the arrival on board of the S. S. Ravenscraig, at Kirkcaldie.") This journal was commenced after Captain Hall's death. Some of the first part of it was copied from another one that was kept during his sickness. It was written day by day as the events happened. The part of it that refers to the boat's cruise after we left our winter-quarters at Lifeboat Cove was written in pencil, and afterward copied into this book. The ship's log was kept by Mr. Chester. It was kept originally in the regular form in the regular log-books. There were two of them. They are now up at Lifeboat Cove, in a water-proof package, wrapped in oil-cloth, and put away in a cairn. We did not bring them with us on account of the size. They were copied into one smaller journal by Mr. Chester during the winter. This is a copy of the whole of the log of the Polaris from the time she left until she grounded at Lifeboat Cove. I believe it is all Mr. Chester's writing.

Mr. Meyer kept Captain Hall's journal up to Disco, I believe, or up to winter quarters. He wrote very little in it himself at Disco. During his sickness he read to me what he had written, between spells.

Mr. Meyer kept it until we got into winter-quarters in Polaris Bay, and after that he took it himself. He kept his journal in a large book. He had three of them. It was put in a large japanned tin box, with all his private papers and all his books of every description, locked up, and set in one of the cupboards in the cabin. I never knew it was gone until we were adrift; that was gone and everything else there, or nearly so. It was there until we were putting the things out on the ice. Mr. Bryan and Mr. Meyer cleared out the cabin ; which of them put it over I don't know. It could not be found when we were adrift; it was put on the ice. Mr. Bryan will know how that was as he put it over. There was a writing-desk, besides; that went too. I don't rightly know what that had in it. I had locked it up. Joe once asked me on the voyage what it had in it; I told him that was something that would have to be settled after we got home. All of Captain Hall's papers were kept in this box with the exception of what might have been in the writing-desk. It was quite a heavy box, and it was quite full of books; the books of ship's accounts were all in there, all his private letters and all his journals of every kind that were aboard, except what was in the desk. They were all missing that night; they were put on the ice, I suppose. I did not see it done, but they were gone. At the time he died his papers were in the room where he did his writing. I saw them there after his death. There were several lying out on a table he had there. He had a small room where he used to do his writing. Several of his papers were piled up under the table. There was no examination made of his papers after his death. They were all put in this box at once, after his death, or a very few days afterwards. Joseph Mauch and Mr. Morton had had charge of all his papers before that.

Question. Were any of his papers burned ?

Answer. At one time during his sickness we were having a talk together about one thing and another. He said he had written a letter to me and took it out. This was after he read his journal to me. He said he had written a letter to me, and he thought I had better not see it; but if I insisted, he would show it to me. I told him it didn't make any odds. He then said he thought it ought to be burned, as he did not approve of it, and he held it to the candle and burned it. I never knew what was in it. He said he thought I had better not see it, and therefore he burned it. This was between his first and last sickness, and during his lucid intervals. No other part of the journal, or anything else, was burned at any time, to my knowledge. Nearly all of his loose papers were put in the box immediately after his death.

Question. Was his journal read about the ship ?

Answer. It was read by Mr. Chester. He asked me to look at it, and I let him have it. I only looked at one statement that Mr. Meyer put in. He read the whole of it to me. The captain had mentioned it to me that Meyer had written it and that he didn't approve of it exactly. All the papers were in the box—every one that I knew of—all his papers of every kind. The small books, I think, (seven in number,) that he brought home with him from the sledge-journey, he gave to his clerk to copy. He was in the act of doing that during Captain Hall's sickness, and it was finished after his death. These contained notes of his whole survey. This is the copy of the seven small books above mentioned. It purports to be a " true copy from the original of Captain Hall's last sledge-party to the north." After Captain Hall's death his clerk had charge of the books, with orders to copy them. He finally had charge of them all the winter we were ashore, and last winter he copied them. They were kept separate from Captain Hall's other papers. Finally six of them were put into a chest where his other books are up there. This one I brought along. There were no other papers left there that were written on the voyage that I know of. If there is any other writing up there I am not aware of it, with the exception of the log-books. There was no formal examination made of Captain Hall's books, papers, and effects after his death. Nobody was appointed to examine them and seal them up, or anything of that kind. The clerk had had charge of them, and had stored them in the box. I never troubled them in any shape or form. I never had occasion to refer to anything, and when I did anything of that kind I generally spoke to him about it, as he knew all about them. All his papers of every kind, except these small books, were put in that box shortly after his death, and remained so until they were put on the ice. I don't think it had been opened for several months. The key was among a lot of keys. I think Hannah had the whole of them. She had control of the keys and about everything Captain Hall had. Whenever I wanted anything out of the trunk I would go to her for it. I found the bundle of keys and gave them to Captain McRitchie, who has them. I never heard of anything being burned except this letter that I spoke of. I don't know what the letter contained. I never saw it. It was at his own option that he did it.

Question. I want to call your attention to the time of Captain Hall's death, and I want to get a distinct expression from you. Have you any reason to believe that Captain Hall died anything but a natural death?

Answer. I really have not.

Question. Did you ever think that he died anything but a natural death?

Answer. I thought there was something very strange about it. I could not believe but what he did die a natural death; but once in awhile, in thinking it over, I thought there was something that appeared rather singular to me; but I have told before what I thought.

Question. Did you ever have any real reason for suspicion? If so, state it.

Answer. No, sir.

Question. Did you ever hear him accuse anybody of poisoning him except when he was delirious?

Answer. No, sir, I think not; and then he accused almost everybody, though he appeared to speak more against the doctor than any one else. We had a very good crew. The mate, the second mate, the seamen, engineers, firemen, cook, and steward did their duty faithfully. I never want to see any better men. I had no occasion to complain of them in any shape whatever after Captain Hall's death or before.

Question. Did you have occasion to complain of anybody else?

Answer. Yes, sir; somewhat.

Question. Let us hear all about it.

Answer. Captain Tyson. He was a man that was rather useless aboard, and complained bitterly about the management generally. He did not appear to be satisfied with anything that was done. I would consult him on a subject and he would perhaps agree to it, and then afterward would say that he thought it was no use to do anything of that kind; that he knew it was of no use. He generally acted in that way. I got so after a while that I did not pay much attention to him. I advised with him very little after awhile. Dr. Bessels and I did not agree very well. Really I could not give any reasons more than I have given in my journal. I proposed a sledge-journey in the summer during the month of May, and got defeated in it entirely. We disagreed on several points in regard to carpenter-work. I believe that was about the whole of it. However, we got along very peaceably, and had no trouble to speak of. We had no outbreaks of any kind.

Question. Was there any disagreement after Captain Hall's death between you and Dr. Bessels about what should be done?

Answer. Not particularly so. It was agreed he should go on with the scientific operations, and that I should take charge of the ship. I assisted him in every way I could in his scientific operations. I had everything done that it was possible to do. He never wanted anything but what was attended to immediately. The discipline of the ship was very good. I had no difficulty whatever with the officers and crew that belonged to the ship. It was not so good as it was before Captain Hall's death. It was felt that discipline would naturally have to relax a little under such circumstances. That is the way I was situated. But there never was any work that ought to have been done but was done readily.

Question. Did you have any difficulty at any time with Captain Hall?

Answer. Yes, sir; twice. We got under way from St. John's about 4 o'clock in the afternoon, just outside of the heads, and by some means or other the steward went ashore and gave the keys of the tea locker, where the tea and other things were stowed in the passage-way, to Captain Hall, or laid them in his room, and we were under way coming out just outside of the heads a short distance, the decks all lumbered up, and getting the anchors onto the bows, and the steward came to me and said, "We will have to break open the locker and get the tea for supper, as we have no tea out." I said, "What is the matter with the tea?" He said he could not find the keys. There was a common staple stuck into the locker. It was not clinched. I did not think what I was doing, nor that there was any harm in it, and drew it out with a marlin-spike, and let the clasp off. Captain Hall felt very much offended at it, and gave me quite a lecture in regard to it. I apologized the best way I could, and finally it was all settled. The next difficulty was occasioned by an unfortunate remark of mine. It was a remark that was very foolish and uncalled for, but it was not intended for his hearing. It was made while in winter-quarters at Polaris Bay. Mr. Chester and Mr. Morton and Captain Tyson and myself had taken a glass of whisky that Captain Hall gave us that morning. We went on deck, and there an argument arose in regard to sledge-journeys. I got rather excited in the argument. I suppose the rest were somewhat excited. Mr. Chester appeared to be somewhat so too, but I cannot say as to that. He is aboard here now. Hays was sweeping off the decks at the time. I will state that Captain Hall had given me orders to save all the chips and shavings that might be around the decks. I went down into the corner of the house and Captain Hall was up over my head. I did not know that he was anywhere about, and I said to Hays, "Save all these shavings and put them in a barrel and they will do—— I might use the expression——

The SECRETARY. Give the expression, if you please.

WITNESS. The expression I made use of was, "They will do for the devilish fools on the sledge-journey." I had no idea Captain Hall was about, and I was thunder-struck when I saw him. That was the only time I ever touched him; it touched him in a very tender point. The remark was not intended for him at all; it was a very useless remark and I was very sorry for it; that was all. It was about the worst thing I could have said in his case, as he was very much in favor of sledge-journeys.

Question. Was there ever any chance to get north with the ship after she got beset in Robeson Channel?

Answer. No, sir; none that I know of.

Question. Was there a consultation there by Captain Hall with Chester, Tyson, and yourself?

Answer. Not with me; I never heard of any.

Question. With anybody?

Answer. No, sir; not that I know of. There was something said after we got into Polaris Bay about the chance to get north. Captain Hall stepped up to the hill himself and looked at the ice and came back and decided that it was impossible. He never asked me about going further at all, but told me that to end the thing he would make that his winter-quarters.

Question. When you first were stopped, and before you floated down to your winter-quarters, was there any lead into the westward?

Answer. We could not find any; we tried, however. That is the way we got beset, in trying to get across to the west side. Once when Captain Hall and Mr. Chester were on a floe, the ice opened a little; we had not steam on then, and if we had had I could not have left without him and without his orders to do so. I think we could have got around that floe, but before they got back again the opening was closed so we could not start over there.

No conversation occurred in which Chester and Tyson expressed a desire to go north while I expressed a disinclination to do so. I never so expressed myself. I have seen that report printed in the papers, but it is not correct. No man in that ship would ever so express himself to Captain Hall and get along with him. I think I should be the last one to undertake to say anything of that kind. I did my very best to get the ship north. I never said anything about never going any further north. I never said that whoever wanted to go north might go, but I would not. I never made use of that expression in my life. I never said so to Captain Hall nor I think to anybody else on board that ship. Captain Hall was a man who would not hear such a phrase uttered by any one such as I have seen reported to have been said. I never expressed myself as being relieved when Captain Hall died. I never made use of such an expression; I thought right the reverse, and I think so still, that I got into more trouble through his death and had a great deal more to contend with twice over than if he had lived. I did make one remark after his death. I was aggravated about something, and I said, while speaking about Captain Hall's death—I do not know how it was brought in—i said, he has got me into a fine scrape and has left me in it. That is all the remark I have any recollection of making after his death regarding his decease. It was very careless in me to make such a remark, but I was a little irritated about something that was going on at the time. I meant by that remark that I had now the whole responsibility of trying to get through with the enterprise the best way I could.

Question. Did you ever say to Henry Hobby, or any one else, "Henry, there's a stone off my heart?"

Answer. I do not recollect of ever saying such a thing, and I do not think I did. I am sure I never did.

Question. While speaking of Captain Hall's death, I mean.

Answer. Yes, sir; I understand. But I never did make use of such an expression. If I did it was foreign to what I felt.

Question. Did you ever say, regarding the journal, or any part of it, that you were glad it was burned or destroyed, as part of it would have been unfavorable to you?

Answer. Never. I never said anything of that kind. All that I ever spoke about was that letter that he burned, to Mr. Chester and Mr. Tyson, and what Captain Hall said when he burned it. There might have been other persons present when I mentioned it to these gentlemen. I do not know how that was. I do not know what was in the letter. I said that he had burned it up and that there might have been something in it against me, or must have been. I do not know what it was; but by the remark he made I thought there must have been something in it that referred to the remark that I had made and which I have repeated here. There was nothing else that I knew of that it could possibly have referred to. If I had not mentioned its destruction to any one, I do not think that any one would have known anything about it. There might have possibly been an Esquimaux man or woman present at the time it took place; but I do not recollect anybody having been in there.

Question. Did you ever have any difficulty with the doctor?

Answer. Only once; I had a few words with him upon one occasion. I had been taking something to drink, and he said something to me regarding it. I just took him by the collar, and told him to mind his own business. That is all the difficulty I ever had with him; that is, openly. That was coming down out of Kennedy Channel, after we had started to come home. It was about taking something to drink; that is all. I went to the aft hatch to get something to drink. He was down there at the time and made some remarks about it. I do not remember what he said, exactly; it was alcohol reduced that I was drinking; alcohol and water, I suppose.

Question. Was not the alcohol put on board for scientific purposes?

Answer. Yes, sir.

Question. What did you drink that for?

Answer. I was sick and down-hearted, and had a bad cold, and I wanted some stimulant—that is, I thought I did; I do not suppose I really did.

Question. Was there any other kind of liquor on board?

Answer. No, sir; not that I know of.

Question. Were you in the habit of drinking alcohol?

Answer. No, sir.

Question. How did it get into the after cabin?

Answer. It was brought up from the fore-peak.

Question. Is that where it was kept?

Answer. It may have been kept in other places.

Question. How was it brought up?

Answer. By myself. There was a half-pint bottle or pint-bottle full; I cannot tell which. It was a very small bottle.

Question. Are you in the habit of drinking?

Answer. I make it a practice to drink but very little. I did take too much twice during this voyage, that I remember: once the latter part of April, and on the occasion I have just referred to. When I so indulged in the latter part of April, it was when we were in winter-quarters. The ship was not moving then. The other time was the night that the ships got beset, coming down Kennedy Channel, the same night that I had the difficulty with the doctor; we were tied fast to the floe. I did not consider, however, that I was not in a condition to do my duty. I merely felt the liquor. I do not think a stranger would have seen it on me at all. I had drank occasionally before, but not to any excess.

The liquors that went on board the ship were under Captain Hall's charge during his life-time, and I had the keys after his death. They were in the magazine. They consisted of wine, whisky, and, I believe, some brandy. I think there was a box of brandy there—alcohol, the last winter, four cans. I took two of them; the other I do not know what the doctor did with. The two I took were taken out onto the ice and their heads knocked in with a pick-ax. That was during last winter, after we had got ashore. We had those four cans left when we got ashore. I do not think it was colder at Polaris Bay than at Lifeboat Cove. There was a great deal more wind in the first winter-quarters, heavier gales than in the last. In the last there was a steady breeze, but no very heavy gales. In the first quarters there were some calm days, but the wind blew very heavily most of the time. There were musk-ox, a few foxes, hares, and lemings at Polaris Bay. There were Brant geese there in summer, the king-duck, a few snipes around the shores, and several land-birds and snow-birds; there were also some gulls. We saw no deer up there. Where we made our second winter-quarters there were deer instead of musk-ox, and more foxes. We found rabbits, also. We saw very few rabbits until we got farther south. When we got there, there were more than we had ever seen before. We saw them on the hill-sides everywhere, and down at Sorfalick, and every island we went to. The men went around there and found them very plentiful in the spring. At Polaris Bay there were some flowers, sorrel, willow, and stunted grass; and finally we raised wheat there. I got it to growing. The head got out of the barrel and some was spilled on the ground, and before we left there it had sprung up two or three inches. This was during the summer. The summer began there about the 1st of June. There were some mild and very pleasant days in May. It began to get very well settled about the 1st of June. The ice did not make as a general thing at that time. It used to make occasionally in very calm days, but it generally went away when the sun came. The aurora borealis was very faint; there was more in the southern portion than there was farther up. We saw shooting-stars very often at our second winter-quarters in the spring. While we were in the boats we saw very few weals out; but we never got a shot at one all the way down that I know of. Our provisions held out well, and even if we had not seen the ship that rescued us I think we could have worked down to Upernavik. I have no reason to doubt it. I think we would have gotten there in about a month or two longer; perhaps in a shorter time. There were two whale-ships still to come, and we might have met the Tigress or the Juniata going up there. This was the 23d of June. I felt confident at the time that we should get down all right. We hadn't heard anything about our comrades that were left on the ice, until we got on the Ravenscraig. None of the natives knew anything about them. I inquired of every one who came to us all the winter through, and no one knew anything about them.

The people on board the Ravenscraig had heard of it at Disco. These papers that came in that box from England are all the papers that were preserved that I know of, except what Dr. Bessels might have. Mr. Bryan and Joseph Mauch have a journal themselves. I believe they lost the astronomical records of the first winter. I think Mr. Bryan said he had none. He lost his journal up to that time on the ice, but Mr. Mauch's journal, I believe, is entire from the beginning; we did not leave any records lying about in the house that I am aware of; nothing that was valuable, at least; we wrapped up all of Captain Hall's printed books and put them into a large chest; also everything that I could

3 P

find that was valuable; the most valuable books were put into a large chest and put away. The logs were put into another cairn on the hill; Mr. Chester put them away in a box and tied them up; I had them tied up and had an oil canvas wrapped around them, and then they were buried in the cairn. The books put into the chest were several charts and books of Captain Hall—printed books, his Arctic works, principally; these were put in a chest and set back one side, with orders to the natives to deliver them up if a ship came; we took away everything we could carry with us in the boats; all the papers and records were carefully preserved; when we left our boats there were several things left on the ice; nothing of this kind, but some of the men's clothing and some cans of meat that we could not carry to the ship.

Question. If you had had the Polaris in good condition would this have been a good summer to have gone on north, as far as you could judge from the appearance of the ice?

Answer. Yes, sir; we could have gotten up Smith's Sound, when we left there, about fifteen miles; the ice went clear across there then.

Question. Have you any means of judging whether you would have been able to get farther north than you got the first year, if you had been able to remain?

Answer. No, sir.

Question. Would Newman's Harbor have been a better place for you to winter than Thank God Harbor?

Answer. Yes, sir.

Question. Would you choose that place if you were going again—if you had to seek winter-quarters in that neighborhood?

Answer. Yes, sir; if I could not get on the west side.

Question. Would you rather winter on the west side?

Answer. I should prefer that.

Question. Why?

Answer. Because there would be a better chance to get north from that side, as far as I know. The land stretches somewhat farther north when you get on that side.

Question. You would choose the western shore to work up on?

Answer. Not with the ship but with sleds. With the ship I would take the east side. The tide sweeps strongly down the west side of this channel and crosses down through Lady Franklin's Bay, and there appears to be heavier ice than there is on the other. The current runs south nearly always; at the turn of the tide we would have a slight set to the northward for perhaps a half an hour or so, as near as my recollection goes, then it would start to the southward again; and sometimes it would not start again to the northward, but would slack and then start off south immediately, or very soon afterward.

Question. In the mid-channel does it ever run to the north?

Answer. Very slightly, sir, as far as I know. I never saw it.

Question. Did you ever have any northerly current after you were out upon Cape Alexander and drifted back to Littleton's Island?

Answer. I never perceived any. It was the force of the wind that drifted us.

Question. Did you discover there any tidal wave from the Pacific? Do you know anything about that?

Answer. I could not say, but I have my doubts about it. This sketch of our journey, north by Mr. Meyer, is tolerably accurate. There are several of the islands off Cape Buckner and Cape George Back that are not correctly placed. Mr. Meyer and the doctor are capable of making a correct chart of the cruise. I don't know who else unless Mr. Bryan.

As we went up Captain Hall said he recognized Cape Constitution. I did not pay much attention to that; I was on the lookout ahead. He spoke to me one day about Cape Constitution, and pointed in that direction. I paid but little attention to it. When we passed Cape Constitution we saw the land all the way up on both sides when it was clear. We had some fogs in 81° 20′. When we passed along here where Kane's open polar sea was laid down we found it to be a sound instead of an open polar sea. The land stretches up a little to the eastward of north above Cape Constitution. The chart of Kane and Hayes is incorrect so far as the position of Cape Constitution is concerned and the lay of the land. The latitude I did not pay particular attention to find out whether it was correct or not. When we were going up Robeson Straits we saw land on both sides about Cape Lieber when it was clear. I think a person being at Cape Lieber could see the land on the other side of the straits. I saw it in the dark days during the winter. I could see land across Cape Lieber and from off the ice.

The straits there I should not call one inch over thirty miles. The degrees of longitude up there are about nine miles. At Cape Lieber, with weather clear enough to see Cape Union on the east coast, it would be impossible not to see the east coast unless there was fog there. I saw Cape Union from the Polaris several times. I saw it every time I undertook to go north. Its position is correctly laid down on the chart, I think. All this land laid down on this chart north of Cape Constitution, up to Repulse Harbor on the east, is a new contribution to geography. I have never seen it down on any chart before. It covers the place laid down in Kane's chart as the open polar sea without any land there at all. I thought I also saw land to the north of Hall's Land, northeast from Repulse Harbor, not laid down on Mr. Meyer's chart. I could not be positive, but I felt sure I saw it to the north, from the northern point made by the Polaris, for I saw the land on the western shore a long distance above Cape Union. I don't think I saw it as far as 84°. I saw beyond Cape Union; that is all. I think Cape Union, from the Polaris's farthest point, was somewhat further from us than what is laid down in Mr. Meyer's chart here. It was not sufficiently clear up there to get an observation. I do not know what is the depth of water in Newman's Bay. In wintering there I would expect to lie at anchor on the north side. I did not go up there myself. I only saw it when I went up with the ship. I never left the ship. Captain Hall said when he went up there and came back that it was his belief that we ought to have been in, and wished we were there. He told me so. I do not know whether he told any one else so or not. He said he wished we were there then. We would have been more likely to have been frozen in there, but we would have been some thirty or forty miles farther north. I think that from the records and the logs that have been preserved, Mr. Meyer, Dr. Bessels, and Mr. Bryan can complete an accurate chart. I have been in those waters a great deal, and have had a great deal of experience in those high latitudes. I consider it possible that the Pole can be reached by this route; but in getting a ship through this channel that is now laid down and back again I think will be a difficult undertaking in some seasons. In some seasons you might get the vessel through so as to get up here on the west shore somewhere and get a harbor, and I think that by proper management you might possibly reach a very high latitude and possibly get to the Pole. I do not think it should be attempted with one ship alone; I should recommend at least three. Place one down upon Littleton's Island, perhaps above, and another one on the western shore as far north as I could get it, into a safe place; with the third I would proceed on as far as possible, having these two to fall back on as a means of escape. With the third ship I would push north, without looking behind, and be prepared to abandon her up there; I should hardly expect to get her out again. A powerful steamer, with good sailing qualities, is what I would prefer. She ought to be a ship of about 130 feet long at least and 450 or 500 tons, on account of carrying sufficient coal. She ought to be 28 or 29 feet of beam; something like the Tigress—I have never seen the Tigress, but I know pretty well what she is—or like the Arctic for instance. If the Arctic were stronger she would be a splendid ship for such a purpose. I refer to Captain Adams's vessel, the one that we came home in. She is a wooden ship, about what would be needed, but it would need of course to strengthen her; she has great power as a steamer. The temperature of the water at Polaris Bay was about 29 in the summer. That was the highest it got, I think. I do not recollect ever hearing of its being 30. I did not pay much attention, however, to that part. Whenever I made any inquiry about it the response was that it was 28 or 29. It would freeze over night when the sun was low. At Lifeboat Cove the character of the bottom where the ship lay was rocky, and between the rocks we would find a little mud, but that was all. It was all a rocky bottom the first winter. As far as I know, the bottom was very soft and muddy. Once in a while there was a large rock. Right where we lay there were very few rocks. I saw the bottom all the way off from the ship a great many times as we were paddling off in clear water. There was some little grass-weed on the bottom, but very little. When we were in our boats and making our escape from Lifeboat Cove I could never tell whether we were helped much or not by the current. We either had a fair wind and were going under sail or else pulling. We pulled with six oars, such as they were, but we did not meet any current against us. If there had been very much of a tide we would have known it.

Examination of Hubbard C. Chester.

My name is Hubbard C. Chester. I am thirty-six years of age. I was first mate of the Polaris. I have been a whaler by profession. I had been in the waters of Hudson's and Behring's Straits before I went on the Polaris expedition. I had never been up to Baffin's Bay before. I went on the Monticello, that took Captain Hall up into Hudson's Bay. I was mate of the Monticello. That was in 1865. Captain Edward Chappel commanded the vessel. I then came home and went round the Pacific. I had made one voyage before with the same men and in the same ship. I went round Cape Horn in the Monticello. I was in Behring's Straits one season. The next season I was mate of the Peru. The next year I was mate of the Daniel Webster, of New Bedford, and the following year of the bark Eagle, also of New Bedford. They were all whalers. In the spring of 1870 I came home from there; from the Sandwich Islands to San Francisco, across the continent to New York, and I think it was the latter part of August, 1870, that I engaged in the expedition with Captain Hall. Captain Hall was looking around for a captain. Captain Chappel was the man who came on here and assisted him in selecting a vessel, &c. He is the man who expected to go. The reason, I believe, that he did not go was on account of his asking too large a salary. After Captain Buddington came home, Captain Hall decided to take him. I sailed with the Polaris from Washington, at this navy-yard, when she went out; from New York on the 29th of June, and New London on the 31st of July. We arrived at St. John's on the 11th of July, I think, and left the 19th. From thence we went to Fiskernaes, where we arrived the 27th of July; thence to Holsteinberg; thence to Disco on the 4th day of August; thence to Upernavik the 21st; thence to Tessuisak, which place we left on the 23d of August. From there we made no stops. We went on from Tessuisak up Baffin's Bay, skirting west of Northumberland Island, up through Smith's Straits into Smith's Sound, keeping over toward the westward and coming up north off Cape Napoleon. We made nearly a straight line across from Cairn Point to Cape Hayes. We met ice off below Cape Hayes and Smith's Sound. We then put to the westward in order to get round the ice, and steered for Cape Hayes. We kept on up, and landed near Cape Hayes. Captain Hall and myself went in a boat to examine the bay and see if it would make a good harbor for the ship in case we were obliged to put back on account of meeting with ice. From that point we kept close to the westward coast. We passed by Cape Constitution and passed through Kennedy's Channel into what was formerly called "Kane's Open Sea." We steamed through that. We saw the land on both sides. We passed Lady Franklin's Bay. We were on the east coast, and passed Lady Franklin's Bay on the west. We passed up through the narrow channel, about fifteen miles wide, which is now called Robeson Channel. In going from Disco up there we were from the 23d to the 31st of August. On the 31st day of August we made our farthest point north. It was pretty well through Robeson Channel—about the center of the channel. Captain Hall told me at that time that he made it 82° 26'. I believe it was afterward ascertained by Mr. Meyers to be 82° 16'. After we passed Cape Frazier we met no ice, steering through open water. I forget the night or the day of the month it was that we passed the small island that is laid out on the chart near the western shore above Cape George Back. We passed that at night when there was a thick fog. We steamed slowly. The next morning we were off the southern cape of the south fiord, as laid down on the chart. That morning at 8 o'clock the fog let up, and we saw this land. We were so far to the east that we saw no opening to the north, and therefore supposed that we were in the bay, the land being all plain in sight. There were quite a number of altitudes taken that morning at 8 o'clock from the point where we found ourselves off the cape—the southern cape of the southern fiord. The fogs came on again, and we lay there until near noon. It then cleared up again. The vessel lay still, and we got a meridian altitude. From thence we steamed up toward the north, and we made the opening which is marked on the chart as the opening of Robeson Channel to the north. We steamed up pretty near the east shore of the channel. Captain Hall tried to land with the boat, I think, twice on the eastern shore of Robeson Channel. On the 31st day of August, 1871, we got to the highest point we made. The steamer was stopped. We could see through the channel, and there was a water-cloud seen—a dense water-cloud—to the north. I mean a cloud that denotes open water. It is a sort of fog that hangs over the water. I think we could have gone farther north from that point. It has always been my impression that we might have gone on. It was my watch below at the time. I heard them sing out to the man at the mast-head, and heard the man at the mast-head sing out there was a lead close to the land on the east shore, and some one called me. I do not recollect who it was, but some one called me and said that Captain Hall wanted to see me on the house. I went up, and when I got there the officers were all there, and the scientific corps. The names of those who were there are Morton, Tyson, Dr. Bessels, Meyer, Mr. Bryan, and Captain Buddington. The vessel was turned round, and she was then headed to the south. Captain Hall said he wanted to get the opinion of the officers as to what it was best to do. It was the opinion of some that there wasn't any prospect of getting any farther. He didn't say so himself. He asked each one his opinion separately. The opinion of Dr. Bessels was, I think, that we had better cross the straits and try to get up on the west shore; and that was the general opinion of the whole party. If we could not get any farther on this side it was thought better to do that than to keep south after we had reached higher latitudes. The idea was, to work up on the west side of the straits ; but in going across that bay, when near the middle of the channel, the vessel was likely to get beset in the ice. I did not go to the mast-head. I only know what I heard the man sing out from the mast-head ; but my opinion was that we had better go on where we were— on the east shore. I don't recollect exactly what I said. I think they came to the conclusion that they could not go any farther on the east shore. Then the opinion of the party was that we should try to cross the straits and get up the west coast if possible. This was the opinion of Dr. Bessels ; the opinion of Mr. Bryan, I think, was the same ; that of Mr. Meyers also. I thought we should try to push up on the east side. I think I told Captain Hall that it would be better to try and push up on that side, and if we couldn't get up there, then cross and try the west side. Tyson's opinion, I think, was the same. We were in favor of going farther north if we could ; if we couldn't, then to go into harbor where we were, if possible. Captain Buddington thought we could not get farther on the east coast, I believe.

The result was we pushed over toward the west shore and got beset in the ice, and drifted to the south, when we should have kept on the east shore where the ice afterward opened. We steamed in toward land on the west, and all the open water there was in the channel, was on the east side of the channel. If we had forced our way on the east shore, even if we had got beset, we would have been sooner liberated than by going into the middle of the channel, or going off on the west shore. The winds were from the northeast when we got beset and were carried down. Before we put off to the middle of the channel Captain Hall tried to make a landing on the shore.

I think that Tyson went with him at the time. That is the place he called Repulse Harbor, because he could not get on shore there. When we got beset in the ice we drifted down to the south. I think it was the third day of September when we got clear. When we got clear we steamed in toward the east coast, into Polaris Bay. It was quite an extensive bay, and what we called Thank God Harbor was formed by what Captain Hall called " Providence Iceberg," on the south side, and a little indentation on the coast on the north side. We did not try to get north at all from there that I knew of. I was at the mast-head from the time we got clear of the ice, steaming into the harbor, and I told him, Captain Hall, that there was a channel of open water along the east coast as far north as I could see.

We steamed in under this headland, and he called me down and I went ashore in the boat with him. That was the first landing. After we had landed there we came off, and we made some soundings and went in with the ship to anchor. My idea was we could have gotten up from there along that coast at that time ; that is my idea. I thought I

could see from the mast-head clear open water beyond Newman's Bay.
Captain Hall wanted to go north as far as he could. I could hardly
tell you what prevented him from going. He was not much used to
navigation, and of course he depended on some one else.

Question. What I want to know is, whether Captain Buddington was
opposed to going any farther north or not?

Answer. I could not say that Captain Buddington was opposed to
going farther north. I do not know that he was, but I think likely if
there had been some one else there as sailing-master the ship would
have gone farther north ; but his idea was, I believe, that we could get
no farther, and therefore the vessel was turned around. We then commenced to steam across the channel, and we got beset and were carried
down. The third day of September, as I have said, we got clear, and
then steamed into the east coast. We then began to land the stuff.

I think it was the next morning that Captain Hall called me on to the
house. I believe Captain Tyson was there at the time. He asked us
our opinion. He asked us what we thought about wintering there.
We told him that we thought if there was a possibility of going up a
few miles farther in the steamer that we ought to do it, and save a great
deal of hard work and labor in taking things over the ice to land. I
told him that every mile that we could get the steamer up so much
labor would be saved. He then concluded that he would go up on to
the high land bordering on the channel on the east side and have a look
up the channel and see how the ice was. He started the day after. He
did not get up on the high land ; it was almost too long a walk, so he
came back. There never was anything said after that about moving
the vessel out, or moving any farther north. The stores and provisions
were landed there. The observatory was built there, and we began to
make the ship snug for winter.

Question. If you had had command of that ship could you have gone
farther north?

Answer. I do not like to say anything of that kind. I should have
tried hard to. I thought I could see considerable open water at the north.
We knew by the water-cloud that there was an open sea of water there.
That was evidence, because we saw all the time we were in the channel,
when it was clear, this dense white cloud to the north. We knew after
we got through this channel that we would be going into a large bay or
sea of some kind. The best chance I saw was at the time we steamed
in after we got out of the ice, when we were beset there. I think we could
have gone up through the channel on the east coast, because the wind was
to the northeast, and all the ice there was in the channel—was in the
middle and on the west shore. Whether I could have done it I can't say,
but I should have tried it if I had had the privilege. We went into
winter-quarters, and on the 17th of September I went away on a sledge
journey for Captain Hall. I went to the eastward ; was gone seven
days, accompanied by Doctor Bessels and the two natives. We went
about twenty-five miles from the vessel on that journey. We got the
first and only musk-ox that was got that fall. We were absent
seven days. This was a hunt. We came home, and on the 10th of
October Captain Hall started off on his sledge-journey. I went with
him. We started with one sledge and fourteen dogs at first, and we
went back from the first encampment after an extra sledge, so as to
divide the load. The traveling was very bad ; the snow was deep and
soft, and Captain Hall, the natives, and myself had to assist in pulling
the sled. We made six encampments. I think we encamped every
night. We stopped in some of them on our road back some two or

three days. We built a hut every night. We went on for six days. The point that we reached was what Captain Hall called "Cape Brevoort," the north cape of Newman's Bay. We went across the bay on the ice about eighteen miles from the mouth of the bay. We did not go any higher than Cape Brevoort with the sleds. We staid there two days. We built up a cairn and buried a cylinder of records in it at Cape Brevoort, near the beach, up the bay from Cape Brevoort headlands some three miles, I think, but near the shore, where it could be seen by any one landing. We found it was impossible to go any farther overland, or go on the ice in the channel, with the sleds, but I traveled over the land from here with Captain Hall, being absent about eight and one-half hours, traveling in what twilight there was at the time. We only had twilight. We were gone eight and one-half hours. We reached the highlands at what is marked down on Meyer's chart as Repulse Harbor. We crossed in that eight hours from Newman's Bay over to the high land at Repulse Harbor, and staid on this high land looking right down from the elevation. We could see the land trending off to the east, on the east shore of Robeson Channel, and turned off rather more rounding than on the Meyer's chart, and a prominent cape off to the east. The land seemed to make to the southward from there, and we could see nothing beyond that cape. On the west side we could see land stretching up, I think, sixty miles that day. It was a very clear day. We stood on high lands, at Repulse Harbor. We could see a cape far on the north, on the west coast, quite sixty miles up. Then there was a dense water-cloud that extended round in a sort of semi-circle. There were places in it lighter than others. It looked like a cloud to me. We came back to the hut, and the next day it was blowing hard, I think. We had encamped there with the intention to go up to the head of Newman's Bay and get on to high land, to see if we could see any more land to the eastward, running off to the north—at the highlands at Repulse Harbor the land was, I should think, somewhere about a thousand feet high. There were hills all along up on that side. We made up our minds that we could not go any higher on that side, and then we started to return. Captain Hall's health seemed to be first-rate. The lowest temperature, I think, that we had while we were away was twenty-three degrees below zero. We were four days in coming home, I think. In returning, we came more on a straight course. When we went, we traveled up the bay, in a ravine most of the way up.

When we reached home, Captain Hall was in good health apparently. When we arrived we saw all hands belonging to the ship. They were banking her in. I went below to clean up, and to look out for our sleeping-bags, and I think we had been in about an hour some one came down into the lower cabin, and said "Captain Hall was sick." I lived in the lower cabin with Captain Buddington, Tyson, Odell, the second engineer, Mr. Morton, and Joe and his family. Captain Hall, Schumann, Mr. Meyer, Dr. Bessels, the cook and steward lived in the upper cabin. The temperature was pretty warm inside the ship. When we went in there that day it was about sixty-five or seventy. I think they kept it up in the cabin about the same, for the order from Captain Hall was to keep the cabins at about sixty-five. I do not remember who it was told me Captain Hall was sick, nor can I remember what they said; some one came into the room and said he was sick, but gave no particular description of the sickness. I went up to see him, I think, somewhere about half past 6 o'clock in the evening. He was lying in his berth. I asked him how he was; he said he felt pretty sick. I think he told me that it was a change of food. He had been eating pemmican, and raw, fat pork, which had disarranged his stomach. He did not say what was the character of his sickness; nor did he say particularly that he had been sick at his stomach. I staid only a few moments with him. I do not know whether it was that evening or the next morning that the doctor told me that his left arm and left side were paralyzed. The next day he appeared to be about the same, I think. The day after he was a little better. I do not know anything about his side being paralyzed. He did not say anything to me about it. Neither did I hear coffee mentioned. In fact, he did not say anything about his sickness; and I know nothing about it, further than what I learned from Dr. Bessels. I do not think I saw him until the next day, when he appeared a little better, I think. That was the third day. He was up and down from that time until, I think, the 6th of November, when he became insensible. I was watching with him that night myself. He appeared to be better than I had seen him when he lay down; but he soon got to breathing pretty hard while he was asleep. I had to call the doctor at such a time; and, it being near the time, I called him, and told him that the captain was breathing pretty hard; and I did not know but what he ought to be waked up. I asked the doctor about it. He said it was all right, and started out as quick as he could to the observatory. He had not been gone but a few minutes before Captain Hall raised up in his berth, and I saw he could not speak. His tongue was swelled. He tried to mutter out something, and I ran out on deck, and one of the men happened to be on the ice taking the tide observations. I sent him right to the observatory for the doctor. I do not know whether the captain spoke after that or not. I have heard that he was quite well; and the next day he was speaking to Captain Buddington and Dr. Bessels, or some one. That I do not know anything about. I never heard him speak. I was watching with him that night before he went to bed. I had been with him before he went to bed about an hour. He seemed to be quite well. Did not take any medicine of any kind that night. I do not think he took any that day at all. I was with him an hour before he went to bed, and he seemed quite well, and took no medicine, and nobody else had been with him; but he went to bed and waked up in this condition that I speak of. There was nothing given to him from the time he appeared quite well until after the time he appeared worse. No medicine was given him, or anything of that kind, that I saw. I was with him every night. The night was divided between Mr. Morton and myself. He was out of his head considerable of the time; indeed, most of the time delirious. He appeared to be suspicious. He was afraid some one wanted to injure him in some way. He was afraid to take medicine of any kind. He was afraid also to eat anything for fear some one wanted to poison him. That was when he was delirious. I never heard him accuse anybody of trying to poison him when I thought he was in his right mind. He accused everybody, I guess, that was in the cabin. I think he accused me. He appeared to be suspicious. If I poured him out a glass of anything, he would want me to taste it first. I did so, but it did not poison me. He thought somebody had guns in the berth there, and he spoke at times of a blue flame he saw coming out of my mouth and the mouths of two or three more persons who were in the cabin. He thought it was poisonous. He thought he saw it coming out of Tyson's mouth, too. He saw it on my coat. He would feel me all over, and try to rub it off.

The doctor attended to him pretty closely. He seemed to do everything he could. I do not know what medicine he gave; nothing more than injection of quinine, I think, into his arm. I saw him do it several

times. He did not give him any other medicine that I saw; nothing
more than a foot-bath and a mustard-bath. The doctor wanted to give
him medicine, but he would not take it. I don't know what he wanted
to give. The captain appeared to be suspicious, and absolutely refused
to take it. Then all the doctor could do was to inject quinine in the skin
of his arm. Before he was taken sick this night I speak of he had not
taken medicine internally for some days. There was one day he ate a
great deal, contrary to the doctor's wishes. He ate sardines and other
canned food. That was, I think, the fourth day of his sickness. He
seemed to have a hearty appetite. The doctor did not want him to eat
the food he was eating. It was the night of the 6th that he woke up
worse, and he died the morning of the 8th. He seemed to be unconscious after that. He lay in his berth with his face down all the
time. His face was flushed, and I noticed a good many sores around
his mouth and at the side of his nose. He breathed heavily; not stertorous breathing, but it appeared as though it were hard for him to draw
his breath. He never was conscious after that, that I saw; they said
he was the next day, I believe. He was talking with Captain Buddington and Dr. Bessels the next day, but I did not see that. I could
not tell you what was the matter with him. The doctor called it
apoplexy, and I take it for granted that that was it.

Question. Have you any reason, in any way, to believe that Captain
Hall died anything but a natural death?
Answer. No, sir.
Question. Do you believe anything else?
Answer. No, sir.
Question. Did you believe anything else at the time?
Answer. No.
Question. Did anybody else express to you any other opinion?
Answer. No, sir; I did not even talk with anybody about it. All this
suspicion of his, and all this talk about his being afraid of being poisoned,
were matters of delirium, when he was out of his head; and that was
so understood at the time. He was buried on the 10th day of November, two days after he died. His grave was dug on the shore, and service was read. I was present at his burial. It was day-time, but it was
all darkness there at that season. Everybody was kind to him while he
was sick, and paid every attention to him they could. Nobody neglected or ill-treated him in any way.
Question. Was the doctor kind to him?
Answer. Yes, sir, and attentive. When he had these outbursts of
suspicion, they tried to pacify him and pass it off.

Captain Buddington took command of the Polaris after Captain Hall's
death. There was no formal assumption of command, but he took command by common consent under instructions.

On the 21st November the ice broke up in the harbor, the ship driving against an iceberg. The third day after that, the ice having got
sufficiently thick over the harbor again, we sawed the vessel out clear
of the iceberg. On the 27th of November a heavy gale from the southwest drove an iceberg in upon the vessel. The tongue of the iceberg
coming under the vessel is all that saved her at that time—keeling her
over so that it broke the ice down on the port side instead of going
through her. It lifted her up. She remained on that berg during the
winter. That is where the vessel received the most damage—from the
rise and fall of the tide. She was over so much that it was uncomfortable living on her. It was almost impossible to get around on deck, the
ship was over so much. We lived along in winter-quarters there all

43

winter, and everything went well. Every assistance was rendered, I believe, in the scientific operations that could have been by the captain and officers at that time.

I do not know what was done with Captain Hall's papers after his death. I saw some of them once or twice within a short time after his death, but I could not tell whether Captain Buddington put them away or not. His writing was peculiar. There was something about it that I could distinguish. I do not know anything about his journal. Do not know who kept his journal for him. I think that Mr. Myer kept the journal from Disco Bay up to the time we went into winter-quarters; and, I think, after that Captain Hall kept it himself. It was kept in one large book similar to the one in which Captain Buddington's journal was kept. I do not know anything about a tin box in which his papers were put.

Question. Was anything done with his effects after his death; were they examined, sealed up, or anything of that kind?

Answer. Not that I know of. At one time I spoke to Captain Buddington, a day or two after his death, and suggested that Captain Hall's papers, &c., should be kept under lock and key. He said he would do it. I never saw any of his things or papers after that time. I never heard anything about any part of his journal being burned or destroyed in any way.

Question. Did you see, after his death, his journal read about the ship?

Answer. I think I saw it in Captain Buddington's room once or twice.

Question. Did year hear anybody after his death say that he felt relieved, or anything of that kind?

Answer. No, sir; I never heard anybody say that there was a load off of his heart, nor anything of that kind. It would have been something that I should have recollected pretty well if I had heard it; but I know that I did not hear anything of the kind. During the latter part of the winter Dr. Bessels and Captain Buddington were not on very good terms, but what the difficulty was between them I do not exactly know. I never heard any words pass between them that winter at all, I never heard either of them say what the difficulty was. I went into the observatory one day, I think in the latter part of February, and spoke to Dr. Bessels, which fact he will likely remember. I told him I did not hardly think they were doing right; that they were the two men that would be looked to to carry on the expedition, that they should consult together and make preparation for spring work. He concluded at that time, I think, that he would write a letter to Captain Buddington, but I cannot say whether he did so or not. I think I said the same to Captain Buddington. I told him that they were not doing right, that I thought they should consult together and make preparation for spring work. What answer he made me I do not now recollect. After that Dr. Bessels made a sledge journey to the south. Robeson Channel being open the most of the winter, we thought if we got north we should have to do it by boat. The 1st of April the boats were taken alongside, and built up on; we could not go north in the ship, because we were frozen in solid; but there was open water in the straits, and moving pack-ice up and down. Along the middle of March I think it was frozen over entirely, and remained so for a month. The ice was moving up and down, would go north and then south again; I cannot tell you what made it go north. We made a number of sledge expeditions to the outer cape, during the winter, to the north of Polaris Bay, and open water was seen near the east shore of the channel. Sometimes the ice would be moving north and sometimes south. The channel seemed to be about fifteen miles wide at the narrowest place. It was darkness at that time, and we could not distinguish water from ice at a certain distance. I do not recollect the date when Dr. Bessels started on this journey south. The date is in the log-book. I think it was somewhere in the latter part of April. I kept the log-book. That was saved all the while. That shows a general statement of all that was done. Several parties were sent out hunting while we were getting the boat ready and waiting for open water in the channel. Dr. Bessels went once; Bryan and the two natives, Joe and Hans, went with him. They were gone, I think fourteen days. I do not know where they went except from what they say themselves. He claimed to have gone about sixty miles, I think, south of our winter harbor. After he came back we were waiting for the opening, and while waiting the scientific operations were carried on all the time; but nothing else was done until we started on that boat journey, except having the provisions got from the shore to the vessel that had been landed the fall before. The ship was leaking; we were pumping with steam all the time trying to keep her clear. We had fire under the small boiler, just enough to work the donkey-pump.

We started on that boat expedition on the 3d day of June. The 5th I left Cape Lupton. On the 6th near the outer cape, I lost my boat and nearly everything I had in her. I was then obliged to go back to the ship over the land and ice. I fitted up the Hagleman canvas boat and left again on the 12th of June. That boat was, I think, 24 feet in length and $4\frac{1}{2}$ feet in breadth, made with canvas stretched over a frame; I think the frame was made of oak. There was nothing put on the canvas at all; it was just ordinary canvas. This boat leaked badly. The first day after I left Cape Lupton I went up twenty-three miles before I landed; she then leaked so that we had to keep one man constantly bailing to keep the water out of the boat; there I found Tyson and his boat's crew. They started from Cape Lupton the third day after I left with the first boat that I lost; here we lay at that place on the edge of the floe about a week. The pack-ice opening a little, we started north again, and reached about two miles and a half from where our first camp was on the ice. That is the farthest point we got with the boat above the outer cape—about twenty-six or twenty-seven miles above Cape Lupton. We returned by the mouth of Newman's Bay, the pack moving down and south all the time. When there were one or two severe gales of wind from the southwest the ice started north a little, and when it was not blowing strong from the southwest the ice was moving south all the time. If I had had my first boat I thought I could have got across the channel, and I should have tried it. The ice opened once or twice. The wind was blowing fresh. The canvas-boat we could not pull; the boat's crew got a little frightened at the condition of the boat. If I had had my first boat I started in, I should have tried to get across, and I think I should have fetched the west shore. The other boat lay there on the ice, a little to the south of where I was, about a mile. We were not able to get any farther north than we did with the boats. I staid over until the 20th of July, I think; then I had to abandon my boats. I received two notes from Captain Buddington to return to the ship. We had to stay there so long that I thought we might get short of provisions. Two of my crew volunteered to go to the ship and bring back some bread. If the ice did open we would not have to return before the latter part of August. When the men returned to the ship the captain kept them there, and sent back a native with a note to me, which I have in my pocket.

The two letters marked respectively "No. 1, C," and "No. 2, C." "No. 1 C" are as follows:

Mr. H. C. CHESTER:

"SIR: Received your letter yesterday, and started north under steam. Have been moving along the pack edge and firing off guns to attract your attention. The present condition of the vessel requires your immediate return. We are going back to the old harbor, where you will follow us with both boats immediately. We have attempted a landing on the cape south of Newman's Bay, but in vain, and have followed along the pack edge to discover a lead without success.

" Yours, respectfully,

"S. O. BUDDINGTON.

" Hans will come back with you."

Captain Buddington sent the boat's crew back afterward, and they brought the second note.

The second note, "No. 2, C," is as follows:

"ON BOARD UNITED STATES STEAMER POLARIS,
" *Thank God Harbor*, *July* 1, 1872.

" H. C. CHESTER:

" SIR: Your presence and that of both boats' crews are required on board, because I intend to get the vessel as far north as possible, and at as early a time as possible. We are burning now from one to one and a half tons of coal daily to keep the vessel free, all the bulk-heads and other spare wood being used up.

With the crew which has remained on board I cannot proceed with sails, and if open water makes north, we can penetrate our way through the ice far better, provided you are aboard, and do not run any risk of getting separated from any one of the party. The ship has been full of water once, and most of the perishable provisions in the hold have been spoiled. It occasioned [happened?] that the limbers had been choked, and the water could not pass the bulk-heads, which was not discovered until fore-peak and main hold were nearly full of water.

" Yours, respectfully,

" S. O. BUDDINGTON.

" N. B.—If, however, you think it advisable, after consulting with Captain Tyson, to proceed farther north with the boats, after having carefully read the above information, I am not the person that will attempt to stop you from doing so.

" S. O. B."

After receiving the first letter, I sent Captain Buddington a note requesting one of my boat's crew to return. The prospects were that we should have to lay there some time to get the boat down by water, and if we had to take it by land I wanted more crew than we had there. Another idea I had was that if there was an opening while I was there, I should have proceeded north, because I was under the impression that when I started overland to go to the ship I had gone as far north as I should go on that expedition, for I knew when we could not work up Robeson Channel with a boat, they could not do it with a ship, especially a leaky one. The next day, I think it was, Tyson wanted to get his boat into the land, and I sent my boat's crew to assist him. They worked two days and one night to get the boat in to land. One of my boat's crew returned, and it was there four or five days on the edge of the ice. There was no opening either way, north or south, and I took my boat in to near Cape Sumner or Sumner Headland. The boats were both left there; one of them secured by canvas being taken from off the frame and the frame folded up and laid on the sled. The instruments and everything we were obliged to leave were put under the canvas and then stones piled on them. The other boat was left right side up, with a boat-cover over her. That boat was stove before we left there, but I secured her with ropes and stones before I left. As quick as I got my boat secured, I sent all my party overland to the ship except Herman Siemans. I kept him there till we got all the clothing dry and packed up snugly to leave, in case we should want to come back and get our boats to go north. We then walked back overland. Mr. Meyer had been with me on this journey, and Dr. Bessels was with Tyson in his boat. Seimans and I were about twelve hours, I think, in walking back overland to the ship. We found the ship at anchor, at the same place, at Thank God Harbor. She was grounded at low tide at the stern. She was leaking, and we were then pumping by steam. Several days after I got back we commenced pumping by hand, and we found that we could keep her free by pumping from five to eight minutes in an hour with a large hand-pump. The pumping by hand was continued up to the time we got up steam to leave Thank God Harbor to go south, which was the 12th of August. On the night of the 11th of August the wife of the Esquimaux Hans had a son born at Thank God Harbor, in latitude 81° 38'. He was named Charles Polaris, after Captain Hall and the ship.

In starting on the 12th of August we steamed down through the ice until we got to the south cape of Polaris Bay; then we came into open water. We steamed down Kennedy Channel that night; we had very little ice. In the morning we had some fog. We were steaming along with full head of steam through the clear water, when, about eight o'clock, the fog lifted and we found we were near a small island, which, from its peculiar shape, we recognized to be the same island we had passed through the fog in coming through Kennedy Channel. We were about five miles from that, and on the other side. We were between this small island and a large one that lies near the middle of Kennedy Channel. That is not marked on the chart. There we were beset thirty-six hours. While there, I think Bryan got some observations on the ice. We again got clear and steamed down farther south until we got beset in the ice in Smith's Sound. I think we never moved from there until we broke adrift, which was on the 16th day of August. We could get no farther; we were blocked up with heavy floes of ice, and were obliged to tie up. We had followed the heavy ice; we tried to keep the west shore, but it was all solid ice, so we moved where the water would lead away from the west shore until we got nearly to the middle of the sound. We were tied up to that floe and floated down with it, until the 15th of October, two months. We made several attempts to stop the leak by drawing a sail under the bow. We also tried to get out some of her ceiling forward and build up a bulk-head to keep the water from flowing aft. The greatest leak appeared to be in the forward end of the vessel. We knew that we had not sufficient coal to pump the vessel to keep her afloat during the winter. We knew we had got to let her sink some time during the winter, even if we had laid to the floe. We were pumping all the time we lay at the floe, the most of the time by steam, not with the small boiler, but with a still smaller boiler, that I suppose we had

had to burn blubber. I believe that is what they call a donkey-engine.
That was rigged so as to work the donkey-pump. It did not consume
much coal. We then drifted through Smith's Straits. We drifted past
Cairn Point two days. There was not much ice to the south, and we
were going with the current pretty fast. On the 14th a heavy gale came
on from the south. It was about 6 o'clock in the evening of the 15th
when the ice first broke around the vessel, setting her off on the starboard side, leaving open water on one side. We still were fast to
the floe and driving with it. We kept on driving with the floe until
we met the ice that first nipped the vessel; she was driven out
on the ice, and there was so much snapping and cracking at first
that I guess there was no one aboard but what thought the bottom
was out of the vessel entirely. Those who were on the ice were very
glad to get there. They considered themselves in the safest place
there; everybody thought that the safest place. I know at that time
it was very difficult to keep men enough on the main-deck to get the
provisions and stores off the ship. We worked until we got the provisions off the main-deck. Then I told the four men, who are here now,
to get out on the ice and begin to drag the stores and provisions back
from the edge of the floe. I then turned to go into the house to get the
ship's log and a clothes-bag of my own to jump out with, and one man,
G. W. Lindquist, started down the ladder, but the ship started so that
he could not get down the ladder. He then went on the ice on one
of the hawsers. He was the only man that went away from the ship
after I told them to get off. There was no other way for them to get
out except on the hawsers. In a moment the ship broke adrift on the
floe. There was a heavy gale at that time; it was dark and there was
a snow-drift. There was a moon, but it did not give much light. We
could not see much in the snow-drift. The ship broke loose, and I saw
the piece of the ice upon which part of the provisions were, brake
adrift at the same time the vessel did, and I saw one or two men on that
piece of ice, but we could not render them any assistance. The first thing
I did as soon as the vessel broke adrift—as I found she was taking water
fast—I got the men out to clear away the snow and get at the deck-
pumps; and all but the firemen and engineer worked at the deck-pumps
until we got the fire going under the small boiler to pump the vessel
with. She drifted to the north and east in clear water. I think it might
have been three-quarters of an hour that we kept her afloat with the
deck-pump before they got sufficient steam to pump with. We were
obliged to do that to keep the water from getting up to the fire; we
were just able to do that. When they got the fire under the smaller
boiler, they were able to keep her clear with the donkey-pump.
We found ourselves at daylight about six or seven miles to the north of
Littleton Island. We were about three miles from the mainland. I
do not know how far we were from where we got adrift. We could see
no land. I think that I saw land once, when the ship was driving away,
but could not say positively. When it came daylight, and it got light
enough for me to go to the mast-head with glasses, I did so; and I saw
a piece of ice with provisions on, that we had landed—or a part of the
provisions. It was about four miles from us, in a heavy stream of ice
that was south of us—between us and Littleton Island. It extended
off to the north and west, across the strait. South of that was an open
sea of water about ten miles in extent; and then I could see the edge
of the main pack of ice south of that. Where I saw this piece of ice
with provisions on was in a narrow stream of heavy ice. I did not see
anything of the floe that we had been tied to. I do not know of any one

else being at the mast-head on this occasion. I believe there was one seaman who went up to the mainmast-head at one time when I was not there. I was up an hour, 1 think, the first time. It was about 6 o'clock in the morning when I was there. Between us and the piece of ice that I saw with provisions on there was nothing but small ice—newly formed ice seven or eight inches in thickness. This piece was not over fifteen yards across. At that time we were pumping with steam on the vessel. I came down from the mast-head and began to clear the lockers that we put up in the fore passage, to make us a boat. We had no boats on the vessel, and I conceived that it would be necessary that we should have some kind of a boat in case any accident should happen to the vessel. We had to get out some way. All hands were set to work making provision for getting out the coal and making the boats, and so on. We at that time had an idea that we would have to get on the ice right where we were. We were bound in the ice. I saw that all around us it was newly formed ice, about eight inches in thickness. While we were at work getting ready, the ice opened in between us and the land, and a light breeze sprung up from the north. We made sail, and with the aid of steam in the smaller boiler, after cutting the ice out of the propeller well, and away from the rudder, so as to move it—and which, of course, took some time—we got started in toward the land. I think it was 4 o'clock in the afternoon when we grounded the vessel as near as we could to the shore, or shore-ice. I think it is the only place within three hundred or four hundred miles either way where the vessel could have been grounded. It was the main point where the ice was clear very late in the fall, and where it was clear early in the spring. There was a strong current setting down between Littleton Island and the mainland that kept this more or less open. It was near what was called, I think, by Dr. Kane, Life-Boat Cove. I was up and down the mast-head all day every ten or fifteen minutes until we got near the land. I went up there to look for our lost parties, but I could not see them at all; they were nowhere to be seen. They were nowhere within twelve or fourteen miles of us, unless they were behind Littleton Island behind a large iceberg that lay outside of it, and close to it; because if they had been I could have seen them from where I was with the glasses I had, from the mast-head. I could have seen them if they had been anywhere within ten miles of the vessel. I did not see the house which we built. I saw nothing but the small piece of ice broken off with the provisions on it. It drifted down not quite so fast as the vessel. I do not know what became of that. We did not make any attempt to follow it, because we could not. There was not any more coal than enough to have got up steam in the large boiler. We had to follow the lead of the ice toward the shore. We could not go any other way. We did not expect when we started to get ashore, but thought that we would get as near the land as we could. We kept on drifting to the south a little all the time with the current and the wind, and we reached the land before we got as far south as Littleton Island. I think if we had seen our comrades on the ice we could have got to them. With the wind the way it was that day we would have tried to have got to them with sails.

I can only account for our not seeing them while they could see us in one way. When we steamed in and got near the land then I was on deck, and no one was at the mast-head. We supposed then there was no possibility of seeing our party anywhere, and the only hope we had was that they were near the land. We knew that they must be near the land on the east shore, and indulged the hope that Hans, who was with the party and was acquainted with the country, and had lived there so many years, would, as quick as daylight came, have them take their boats and try to reach the land with the party. If they saw the vessel at all it was just before she struck the shore after she got inside of the range of Littleton's Island from them. The time that they saw us must have been about the time that we were just reaching land, and at that time there was nobody at the mast-head. We had been looking for them all day, but had given up all hopes of seeing them. I was at the mast-head of our ship all through the day until just before the vessel was grounded. We had good glasses, and I could raise nothing that looked like boats, men, or anything of the kind on the ice. All I saw was this piece of ice with provisions on. Where we grounded was about two miles or two miles and a half northeast of Littleton's Island. If they saw the smoke-stack they must have been north of Littleton's Island; because I have been to Littleton's Island since, and I could not see the smoke-stack from the ice at Littleton's Island, and that was only two and a half miles from where the vessel was run ashore. There were hummocks and small icebergs that lay to the south of us, between us and along on the shore, the point that made out toward Littleton's Island. The only way we can account for not seeing them is that they must have been behind Littleton's Island, from us, or behind the berg that was there, because from the south part of Littleton's Island was all open water, which extended across the straits. It was several miles south of Littleton's Island, to the edge of the main pack that extended the whole width of Smith's Straits. They might have been behind some of the hummocks, but I think I could have seen seals six or seven miles distant on the ice that morning, for it was clear, fine weather. They must have been behind some obstacle, because there were nineteen people, including the children, two boats, India-rubber blankets, colors, the house, and all the provisions, and that would make a pretty extensive object. It is possible there might have been refraction in the atmosphere, such as frequently occurs at sea, which would have lifted the vessel up by a mirage, which brought the vessel in sight above, while we could not see them, but I did not know that there was any such thing, and did not notice anything of that kind in looking toward them. Northumberland Island is distant from Littleton's Island about eighty miles. I will state that Captain Buddington was on the house all the time, and nearly all the hands were on deck. If we had rescued the party on the ice, we would have been able to have recovered our ground better. They would have brought the boats which we needed, but we should have had to build other boats, because those they had would not carry the whole party.

I will state that, as regards personal safety, I think I should have preferred being on the floe to being on the ship, because we did not know the condition the ship was in at the time of the separation. The snap and crack of the timbers of the vessel when she was nipped and thrown on to the ice of course led every one to feel uneasy. There was no one on board but who thought that she was more or less injured, and when she settled back into the water, that she would likely fall to pieces and sink. That was the general impression of all hands at the time, I guess. The other party had the boats and the kyaks, the natives, and the scow ; and most of the provisions on the vessel were landed there. All the skins of the musk-ox and the largest part of the clothing of all descriptions were hove out on the ice. I do not think Captain Buddington ordered any men to go on to the ice. The only order I heard given was to " overboard provisions." About the first thing we did was to lower the boats. The most of the men had to get out to take these boats clear of the side of the ship. But before doing that most of them threw their clothes-bags out and got them on to the ice. After they got out there they didn't care about coming back on board the ship again, and remained on the ice. Of course it was necessary to have some men there to take back the provisions. I think Captain Buddington ordered the boats to be lowered. When the ship grounded at Lifeboat Cove we got out lines and made fast to the hummocks of ice there. The next morning at daylight we got up what coal there was in the bunkers on deck. The next day we sent down all the topmasts, booms, and gaffs, and dragged them on shore and built the house, and then we next commenced landing the provisions and taking off the coal that was on the vessel. The stock of coal that was left on the steamer was about five and a half tons. We built up a house and were there some three days, I think, before we got the house finished and got moved in. The fires were let go out, I think, at 6 o'clock in the evening. The next morning at 8 o'clock the water was within two feet of the main-deck of the Polaris. I did not examine her condition, any more than I could see that her stem was stove in about three feet, and the stem itself gone, and the wood ends and some of the planks four or five feet in length broken off and turned right around, and some of them were still hanging by the slivers. I do not know whether that was done on the iceberg or not. I do not think it was gone at Polaris Bay, because I should have seen then. It was not gone until we went adrift that night, I think.

I did not know how the vessel could float when I looked at her stem; she was in such a condition that she could not possibly have been repaired and brought out. The stem was entirely gone. Perhaps if we had been in open water, and had plenty of coal on the steamer, so that we could have pumped by steam and kept the vessel steaming, we might have got her into one of the ports of Greenland, but she would not have been safe to have left a Greenland port to have come here in. We could not have kept her up after we had got her in port without pumping all the time. When she had reached the land, she had done all that she could do, and that was an end of her usefulness.

After we grounded we were on the lookout for a number of days, thinking our lost party would land somewhere to the south, and work up to the north. We knew Hans was well acquainted with the country, and we thought it likely that under his guidance they might reach us. We made this place our winter-quarters, and remained until the 3rd day of the next June. We built boats from the linings of the cabins of the Polaris. I superintended the operations of the building of the boats myself. We built two boats for ourselves. We built a small one for the natives there. The ship afterward sunk. As the ice broke up, she worked off shore a little all the time. She was full of water and working off all the time. Her rail was just out of water at high-water mark when we left. But there were lines fast to the shore. One line was let go when we left with the boats. We had to let that go, in order to get by with our boats. I believe we told the natives to make that line fast again.

During the winter nothing of special consequence happened. The scientific observations were kept up. We did not keep up the observations of the tides, because we could not. Dr. Bessels tried to make some arrangement for taking the tide observations there, but he could not do it. He had to go off shore too far, and could rig no apparatus.

I kept a journal until I lost it in my boats in the spring of 1872, when we were up in Newman's Bay. I have kept none since then. I kept the regular log of the ship. That was kept in two of those large books—printed Navy log-books—which had been supplied to the ship. They are at present up at Lifeboat Cove. I made a fair copy from those two books into a smaller book. This copy was word for word. This book is here. We found that these large books were rather too heavy to carry with us in our smaller boats. We found it better to copy them in something lighter, in order that we might save the contents if we could. I therefore copied the contents myself into this smaller book. This copy is all in my handwriting. The original log was all copied by me, and is in my handwriting, except when I was absent in the boat journey at Newman's Bay. This copy was made by myself. This book I have had in my custody all the time. We left with boats on the 3d June and boated down most of the way in open water, keeping the land-floe of ice until the 23d day of June, and about twenty-five miles south of Cape York—Cape York was plainly in sight—we were taken up by the Ravenscraig whaler of Dundee. We were on board of her altogether until the 6th day of July. Then seven of us were transferred to the Arctic. Afterwards, at what time I do not know, but some time afterward, three others were transferred from the Ravenscraig on board the Intrepid, and those are the three that have not arrived. They are Mr. Bryan, Joseph M. Mauch, and John W. Booth. This transfer was made in order that we might be divided up, as all being in one vessel was rather more than it was supposed the stores of one vessel could supply. When the Arctic was ready for home we saw the Ravenscraig and took off the men from her. The other ship, Intrepid, was in sight, steaming away. We had to come away and leave that party. The vessel with the other party is likely in Dundee now, or on her way there. We were treated very kindly indeed. We were taken to Dundee, and there we were cared for by the United States consul and supplied with clothing, and came home in the City of Antwerp. We were in Dundee about four days.

The discipline of the ship was first rate during Captain Hall's life-time. Afterward the discipline the first winter was very good. I do not know but what it was good enough all the time. I do not recollect of ever giving a man an order on the ship but what it was executed very promptly and quickly, without any hesitation, from the time we left Washington City. It was as good discipline as ever was observed on a whaling-vessel. We had a remarkably good crew, as good a crew, I think, as ever went into the arctic regions. They were just the men needed on an expedition of that kind. I do not know whether Captain Hall's papers were put out on the ice at the time of breaking loose. Captain Buddington was superintending all that, and worked himself there. I was at work on the main-deck. If they had been put out, they would have been put out at the stern. Nearly all the provisions were carried back and put into one pile. There were some men that were clearing away from the forward gangway, and some aft. As a whaling commander Captain Buddington, I think, does very well, but not so good for a north-pole expedition. He has not that enthusiasm for the north pole that Captain Hall had, or Kane had. He drank a little occasionally, and I have seen him once or twice in a condition that we would call "boozy." I do not know anything about his drinking alcohol on board the ship. I have seen him boozy when I thought there was nothing else on board; but I do not know anything about his drinking it. I think he had been drinking a little the night we got beset in the middle of the channel coming down. I never heard any words pass between him and Captain Hall at all. I heard there was a little trouble in getting out of St. John's. Captain Hall appeared to have a kindly feeling for Captain Buddington—more than Captain Buddington seemed to have for Captain Hall. I got that impression from what I saw on the vessel of the actions of the two men. He at times rather depreciated Captain Hall, in using language around the main-deck that should not have been used by a man in his capacity. When I say "main-deck," I mean among the seamen. He did this when he was sober. He did not speak very respectfully of the commander, or of the expedition. I cannot, however, recollect any particular words or any particular expressions that he made use of at any time. His idea was, as it struck me, that the enterprise was all "d——n nonsense." He did not seem to have, either, any regard for the scientific work; he thought that was all nonsense too. He never appeared to have any trouble with it until after Captain Hall died, then there appeared to be some little trouble between him and Dr. Bessels. I never heard any words between them. I do not know whether they ever had any or not. They did not in my presence or hearing Captain Buddington expressed himself as being of no use in the expedition, and depreciated Captain Hall in the presence of the men. I do not know that I ever heard him say anything against Captain Hall's authority in the presence of the men. He did not seem to question that at all. I do not know that I ever heard him say that he was no seaman, or anything of that kind, but he regarded the whole thing as foolishness. I heard nobody else make such a remark. I never heard a man on the vessel say anything but what was encouraging of the expedition except Captain Buddington. What I did hear him say in the presence of the men I regarded as very improper, when said by a person acting in the capacity that he was. It was said so that all of us could hear it. It was not especially addressed to the men, but they all heard it. The Polaris began to leak in Thank God Harbor after we got into collision with the iceberg. The next spring, as soon as the water began to make around the vessel, we calked her from the outside. It was at the edge of the water. The ice was making between the sides of the vessel as it always does. Everything was done that a seaman could do with the means at our hand at that time to stop the leak. At subsequent times when leaks or other accident happened, everything was done that seamen could do, or ought or might do, with the means at our command, to remedy these things. Captain Buddington generally gave the orders, and I had the orders executed. I had nothing to do with the navigation of the ship. While I was away on boat journeys I made some observation of the latitude, and I made some observations on the ship. When we left the house, we left behind in it a few cans of dried potatoes and a very few cans of meat; I think there was a little meal in the barrel and a little flour in one barrel, and some bread; we gave them to the Esquimaux. We left no books or valuable papers, nothing but what were put into chests and boxes and stones piled up over them. These cairns were about one hundred yards up the hill, and about twenty feet, I guess, above the sea-level. We explained to the Esquimaux that these were books and papers, and nothing to eat, and told them not to disturb them. There was nothing of value of any description left there that we could take away. The pendulum, the transit, and other instruments, Captain Hall's arctic library and other books, were packed up and left in the same cairn. Dr. Bessels had a trunk with thermometers and some of his scientific instruments in it. They were all put in this cairn. The log-book of the ship was also placed in it. The canvas boat was not good for anything. It would ferry us across a river. It would stand quite a little sea, but then the canvas ought to be prepared so that the water will not go through it. This canvas had been lying out all winter exposed to the weather and the driving snow. It had been on a pile of stones ashore all winter. It was taken right off and put on to a frame, and it leaked pretty badly. When we left the Polaris she was still aground and full of water, and tied up to the shore. I have heard it said that Captain Buddington gave her to the Esquimaux, but whether he did or not I do not know. I do not know whether I heard him say so or some one else. The ice drifted north and south in Robeson Channel both ways. It drifted northward when there was not a south wind or southwest wind blowing. In the winter the farthest we could get out was about this outer cape. Here (indicating on the chart) the ice sets up and down with the current and sweeps up this way, (indicating.) The ice ran down along the harbor. The ice was coming down southward continually right through while we lay at Newman's Bay on the land-ice. The pack was moving south most all the time. At Thank God Harbor there was plenty of open water, still these straits at Newman's Bay were full of pack-ice moving down. The vessel started out from Thank God Harbor two or three times. She came up around the cape part of the way to Newman's Bay from her anchorage here, (indicating on the chart.) She struck this moving ice, and followed the edge of it nearly two-thirds of the way across Robeson Channel—a solid pack edge. The pack-ice went into Lady Franklin Bay. Here by these islands, (indicating,) when we came down, we found a great deal of open water. There was no difficulty in steaming down at all, notwithstanding all the ice that moved through Robeson's Channel while we lay there, which was about forty days. I could not tell which direction the tide came from, whether from the south or north, on the flood-tide; it just rose and fell. I could not tell anything about the drift; I only know this ice was going south all the time unless there was a south wind, and then it would move slowly to the north in Robeson's Channel. I noticed, some days when it was calm, that the ice was moving south over one tide, whether it was flood-tide or ebb; did not see ice disappearing down this southern fiord; it was frozen; at least it was full of ice here when we came out. We found wood on the south side of Newman's Bay, but on the north shore of Polaris Bay we found no wood, nor on the north side of Newman's Bay. It was the same kind of timber I have seen in Behring's Straits. It looked similar to it.

Without concluding the examination of witness, the commission adjourned until to-morrow morning at 11 o'clock.

The chart made by Mr. Meyer is generally correct. There are some small inaccuracies. Cape Constitution is in latitude about 80° 20′, I should think. I think it is about right on this chart. I think there are some inaccuracies in the outline of the coast at Newman's Bay and above. The track of the first journey by Captain Hall and myself is not accurately laid down, but generally the chart is pretty correct in regard to our new discoveries and the coast-line below.

Examination of William Morton.

I was born in Ireland. I have lived in this country thirty-one years. I reside in Jersey City, N. J. I am a seaman—follow the sea for my living. This is my third trip to the arctic regions. I went first with Captain De Haven, in 1850, in search of Sir John Franklin. The second time I went with Dr. Kane in 1853, in search of Franklin, taking another route. This is my third expedition. I have spent most of my time since I came to this country in the regular naval service of the United States, generally as a petty officer—principally yeoman. I sailed from Washington on board the Polaris, as second mate, on the 10th of June, 1871, to New York, with Captain Hall; thence to New London; thence to Saint John's, and from there to Fiscanaes.

Nothing of interest happened up to that time. We went from Fiscanaes to Holsteinberg. Captain Hall thought the Congress might call in there. After a few days we went from there to Lively, on the island of Disco. There we remained a few days, when the Congress arrived. At Disco there were a few words of misunderstanding between Captain Hall and, I understood, the scientific officers—Mr. Meyer and Dr. Bessels. It was, however, all arranged amicably before the Congress left. Captain Davenport came on board and gave advice to the officers and ship's company. From Disco we sailed to Upernavik. I do not know the date of sailing, as I did not keep any journal. From Upernavik we sailed to Tessuisak, which is the northernmost Danish settlement of any account. We went there to get the rest of our dogs and furs which we could not procure in the southern settlements. From there we went through Melville Bay, and made our way north. We left Tessuisak on or about the 24th of August. We went through Melville Bay without any obstruction, except merely taking an irregular route, but we did not meet any ice to hinder us—none, at least, that we could not easily get around. We were at Cape Alexander the third day, almost to the hour, from leaving Tessuisak. We found the entrance into Smith's Strait free from ice, and passed Littleton Island, and there saw a good number of walruses playing. We fired a few shots at them, but without effect. We went up considerably farther, but not so high as Kane's winter-quarters, when we struck off to the west shore, not following the east shore as he did. During that night, about 12 o'clock, we fell in with a barrier of ice that gave us the thought that the passage of our vessel was obstructed in Smith's Sound. We discovered, however, a lead inshore between this heavy floe and the west shore, and by going back on our route several miles we headed a tongue of ice and got into an open lead, and went on without obstruction to Cape Frazer. We passed several known places, but I think it was there that Captain Hall stopped and went ashore in order to leave a depot of provisions, where we could seek a harbor in case of necessity. He found the place too shoal for the ship to rest in to make winter-quarters of, and so we went from there to Kennedy Channel, still unobstructed by ice. We went through Kennedy Channel, meeting occasionally a patch of ice, but not enough to obstruct the vessel from proceeding. We passed Cape Constitution, and recognized it by the two islands, but were not as near to it as I should like to be to make an accurate survey of it with the eye.

Another island that Kane's party did not discover before is on the opposite shore and a little higher north. From the position of Kane's party at Cape Constitution it was land-locked or lapped in with the opposite shore, and was taken for a head land of the main-land. That is about the narrowest place, in my opinion, between the islands. It does not look so wide as it actually is. For instance, Franklin Island from the pitch of Cape Constitution is six or eight miles, but you think, by being in the middle of the channel, that it is leaning right up against the land; and then the other island, over on the west side, is twelve or fourteen miles at least from the shore, though it seems much nearer, and that leaves the channel there, in my opinion, between twenty-six and thirty miles wide, that is, from main-land to main-land. Above that there is an open area of water. Hans and I, when with Kane's party, saw that. We could not see the land to the eastward of Cape Constitution, but, looking westward, we saw land until it dwindled into space some forty or fifty miles off, I suppose. I could not say whether we went between the islands or not, because it may have been my watch below. We went right ahead, and with very little obstruction. In fact, when we got into this open area, the water which we supposed to be a sea we found to be a large bay, perfectly free from ice. This, which had formerly been called Kane's Open Polar Sea, we found to be a large bay, at that time clear of ice. We could see the land on either shore as we passed through, but could not see the land ahead until we got clear up and the fog then existing had cleared away. Then we found an entrance to the eastward. We passed that and a large entrance to the westward, and that is called Lady Franklin's Bay. The entrance to the eastward was afterward called the Southern Fjiord. That is the name given by our party. The entrance to the westward was Lady Franklin's Bay. Steaming across the head of this bay we discovered another channel leading to the north-northeastward, or thereabouts. I should judge it from twenty-six to thirty miles wide and narrower than parts of Kennedy Channel. That was named, by Captain Hall, Robeson Channel, after the present Secretary of the Navy. We went up that channel considerably, I disremember exactly how many miles, and the first real obstruction we met up there was the heavy pack-ice that extended from shore to shore of this channel, with a small lead on either shore. At a place on the east shore Captain Hall went ashore in a boat, on two occasions, to look for a harbor, but found none to suit. He called it Repulse Harbor. The second time he came back and called a consultation of his officers, on top of the house, comprising Captain Budding-ton, Chester, Tyson, Dr. Bessels, and myself.

If there was any other consultation among the officers I was not present, and am not aware of it. Some of these officers were for going north if possible, and others were for looking for a harbor immediately ; and I think Captain Buddington preferred going back, at least, to what was afterward known as Newman's Bay, for a harbor. Captain Budding-ton was in favor of falling back to that place. We tied to the ice at the time, and after a little while we proceeded toward the west shore, where there appeared to be some open water, and possibly a lead along it into an open space of water that we could see in fact; we saw the clouds over it, and it widened where the land fell off on both sides. While going over we got beset, and the ship got nipped, but not to injure her; that is, the ice closed on to her, and she was in danger of being injured. Captain Hall ordered provisions out on the ice, so that in case of accident we might have something with which to support ourselves. Afterward the ice eased off, and the next day we took the provisions on board again. We were then drifting rapidly with the ice down Robeson Channel to the southward again. We reached our highest point August 30, 1871, when this consultation was held above Newman's Bay. Our latitude at that time, by dead reckoning, was 82° 26′, but it was afterwards found by observation to be 82° 16′. I think that was the highest point we reached, and that was the same day that we had the consultation. The next day we were south of that latitude ; we never got any higher than that in the ship, nor did anybody get any higher on land. Repulse Harbor, the place Captain Hall went ashore, was the next highest point we reached ; that is but a very short distance below the highest point. After we got beset, we floated down to the southward to where Robeson Channel widens into the bay. In the bay now called Polaris Bay we got a lead to the southeast, and went into that and got under the lee of the shore in Polaris Bay, some four miles from Cape Lupton, at the mouth of Robeson Channel, and came to anchor there in a kind of cove; it could not be called a bay; it was a sort of indentation in the shore. We came to anchor there inside of a grounded iceberg; we left there the day following. While Captain Hall was on shore he thought he saw a place deeper in the bay, and we got under way and tried to get to it, and after steaming around a few hours we did not find any better place, and returned and came to anchor in the same place inside of this grounded iceberg, which was named by Captain Hall Providence Iceberg. It was grounded in about thirteen fathoms of water. Here we intended to remain, so far as I know, and in a few days commenced landing our provisions on shore.

No attempt was made to go farther north ; it was late in the season. It was dangerous, in fact, and I did not know that Captain Hall contemplated leaving there to go north ; I never heard any suggestion of that kind ; it was beyond the time for navigation. Hard frosts had set in, and we could not have got the boats through the ice, and it was not strong enough to walk on, so we were detained a few days until we were finally able to walk on the ice, and after that we took the rest of our provisions on shore and built a house for observatory purposes. The ship was a full quarter of a mile from the shore, and the house was about a hundred yards up the side of a hill, where our provisions were put. We landed all our provisions there and made preparations to winter, by clearing the ship of almost everything in her ; we cleaned her right out, with the exception of a few trifles, such as whaling-gear and marlin-spikes, which were kept in a store-room on board. We then covered our vessel with canvas made in Washington for the purpose before we left, and made everything comfortable 'for winter-quarters. We cleared out the after lower cabin for sleeping-apartments. The rooms on deck had to be forsaken, as they could not be kept warm. We housed our vessel, and continued there during the winter.

Captain Hall went off on a sledge-journey about the 10th of October ; he was absent some fourteen days. He was accompanied by Chester, the first mate, and the two Esquimaux dog-drivers or hunters. He returned in fourteen days exactly.

Question. How long was he gone ?

Answer. Fourteen days. He came back on the 24th in good spirits.

Question. What time in the day did he get back ?

Answer. It was before our dinner-hour in the afternoon ; I think it was about 2 o'clock, though I will not say as to the hour. I was ashore when he came. I met him on the ice between the ship and the shore. I shook hands with him ; asked him how he was ; he said he was right well, and glad to find everything so well and pleasant on board ; very much pleased with the proceedings since his departure. I went on board with him to the upper cabin, and I staid with him at that time, except when he ordered the steward to get him a cup of coffee. While the steward was gone for the coffee I went to get him a shift of fresh clothing. He ordered the steward to bring him a cup of coffee, as I have said, and he went to the galley and got it.

Question. Did the steward bring it back while you were there ?

Answer. I don't recollect. I went to Captain Hall's private store-room to get him some clothing, and when I came back he was vomiting. I was alarmed and asked him what was the matter. He said, "Nothing at all—a foul stomach." I was not gone more than twenty minutes; it could not be much more. I sought some clothing that he wished to put on.

Question. Who was with him when you went after the clothing?

Answer. Hannah was there, and I don't know whether Captain Buddington was there or not. He came on board also with Captain Hall. There was also Joe, the Esquimaux, and the steward. I don't know of anybody else, except, perhaps, Dr. Bessels.

Question. Was anybody with him when you came back with the clothing?

Answer. Not that I recollect. Other people may have been with him previous to that, but they had gone out; for instance, Chester and Tyson had gone out and shaken hands with him.

Question. Then nobody was in the cabin with him when you came back?

Answer. Not that I recollect now. When I came back I asked him what was the matter; he said there was nothing the matter except a foul stomach. I proposed getting some hot water to bathe his feet, which was done, and his clothing shifted. After we got a clean shift of clothing on him he went to bed. He was then proposing to start the next day on a journey south, and intended to take Captain Tyson with him, but his sickness got worse. The next morning he was so bad that Mr. Chester and myself proposed not to leave him alone during the night. He was alone without any watcher the first night, but he got so bad the next day that after that Chester and myself kept watch with him during the night, watch and watch. Captain Hall spoke against it, and said he did not wish to put us to so much trouble. We insisted on it, and continued it till he died. I heard him asking for an emetic; he said it would do him good. The doctor was there also, at the time he was vomiting and sick, and I believe while he was taking the coffee. He asked the doctor for an emetic, and, as far as I could understand, the doctor said "No," he was not strong enough, or it would weaken him too much, or something to that effect. He got delirious very soon after the second day. He got suspicious of some people, and said they wished to harm him, and he said to me, "They are poisoning me." I thought he was out of his head; indeed, I knew he was. He said to me, "Whatever I want you will get for me, and see that it is all right—see that there is nothing in it. You were a friend of Kane's, and I want you to be a friend of mine." He got me to make tamarind water and 'arrowroot for him. Other things the cook cooked and Hannah administered. But during my attendance upon him he would take hold of my hand when we were alone, and would say, "They are poisoning me, and you won't leave me." On these occasions I considered him out of his head. He was out of his head the most of the time. He continued this way six or seven days, and he then got right smart, and got up. He sat up, in fact, a great deal, on a lounge or bed. He used to rest himself on the lounge, and turn in occasionally. He got up and spoke about his journey, and went about his ordinary business for a day or so, and then relapsed. He then went to bed again, and got worse and worse until he died. The doctor told me, I think the second day, that Captain Hall's illness was very serious, and that he would not recover. That was the day after he was taken, or the third day at the furthest. I cannot rightly recollect what the doctor said was the matter with him; apoplexy, I think. He was not smart in his movements like, but I did not know particularly that one side was affected more than the other. He was feeble like—prostrated. He showed that feebleness very soon; not immediately after his vomiting, but I noticed it the next day, when I put on his clothing. I had to help him, he was so sick and enfeebled at the time from vomiting. He had been vomiting and retching violently for probably ten minutes. He was vomiting while I was absent, and I cannot say how long. I assisted in putting on the clothes. He had my assistance, but he might possibly have been able to put them on himself. I was of great assistance to him, facilitating his movements. While he was sick I was with him a great deal during the day, and generally half the night. Either Chester or I kept watch all the time. Hannah was there during the day-time nearly all the time administering to his wants. After he grew delirious he got suspicious. I never heard of him being suspicious before he got delirious. I understood that he was afraid of almost everybody. Captain Buddington, Dr. Bessels, and even at one time Mr. Chester—the best friend, in fact, he had aboard—he was afraid were going to do something to him. I do not know that he was afraid or even spoke of Hannah and Joe in his delirious moments. He never seemed to be afraid of me before my face. He always thought he could depend altogether on me; but dear knows I don't know what he said when I was not present. He may have said I was going to kill him as well as anybody else, for all I know. He said somebody "had a gun over there." There was no gun there. I hear of his thinking he saw a sort of blue gas coming out of people's mouths. He never struggled with me when I tried to assist him. I heard him struggle with others. I heard Captain Buddington trying to put him in bed when he wanted to go out. I was in the lower cabin, turned in, when I heard it. When I had waked and turned out things were quiet. I heard it, but I did not see it.

Question. Had he taken any medicine, or anything, before the vomiting?

Answer. No, sir; nothing but ˙the coffee which the steward brought him from the cook's galley.

Question. Who gave him his medicine generally?

Answer. Dr. Bessels. I never gave him any. I don't know whether Captain Buddington did or not. I think he did, because he appeared to take it from him. He was opposed to taking medicine from Dr. Bessels when he was delirious. I do not think he took much medicine. He was apparently better for two or three days. He seemed very smart, indeed, and we all thought he was better and going to be the same as usual, and would be able to take the journey which he contemplated to the southward in a day or two. I think he ate some cooked hare that day. I think the doctor objected to him eating so much as he would wish to; but he did eat a good deal for a man that was so enfeebled and sick; for instance, he ate a thigh and leg of a hare, or something like that. I was not present when he was taken sick the second time and had his relapse. That was at night, but it was not my watch. Mr. Chester must have been with him then.

Question. Did he take any more medicine the day that he appeared to be well?

Answer. No, sir; not that I know of. I believe he stopped taking medicine. .I think these expressions of suspicion and distrust of various people were the expressions of a man in delirium, and I have no cause to think otherwise. He never spoke of them in his sane moments to me, or anybody that I know of.

Question. Have you any reason to suppose that there was any foul play toward him?

Answer. I have not, indeed.

Question. Did you think so at the time?

Answer. I did not; it never struck me.

Question. Do you think so now?

Answer. I do not.

Question. Then you consider these expressions of suspicion by Captain Hall the ravings or hallucinations of a man out of his head?

Answer. I do, sir, and I hope so.

Question. Have you any reasons to believe otherwise? If so, state them.

Answer. No, sir; I have not. I have no suspicion to the contrary, and never had, except the reports that I have heard around. I never formed one myself, and never had one. I never had any reason for suspicion or doubt. Dr. Bessels was as kind to him as anybody I ever knew, while attending to him, and administered, I suppose, to the best of his ability, and I saw no reason to suspect or distrust him. I was the only one that was present when he breathed his last. He was in a heavy sleep as I thought, lying with the side of his face on the pillow, his mouth and side of his face down in the pillow. I sat by his side, and he breathed very heavy, and Mr. Chester remarked to me, "He is asleep, and I don't think he is any better; he is very bad." Chester turned in; and after a while I spoke to him, but he made me no answer. I raised his head with my hands, and I saw something about his mouth—saliva about his mouth. I then turned him partially on his back, and put his head a little more upright, wiped his mouth, and put a teaspoonful of some kind of drink between his lips, but he never noticed it. I don't think he swallowed. I had to wipe off the saliva and clean the side of his mouth then. He remained in that position then for some time, breathing shorter all the time, and finally I had to listen to him. At about 20 or 25 minutes past 2 o'clock, when I was with him, he ceased breathing. I kept my cheek close to him, but I could not hear any breathing. I went immediately and shook the doctor and woke him, and told him the captain was dead. I had to call him twice, and he could not comprehend thoroughly. I said, Captain Hall is dead. He jumped out, and I then went down to the lower cabin and called Captain Buddington, and told him the captain was dead. Afterward Buddington called Tyson and Chester and the rest. Chester and all hands were turned in below but myself at the time. There were six or eight people in the upper cabin, but they were all asleep. When they came up Captain Hall was dead a minute or so. While he was in these last moments his face was very placid. There were no contortions; nor was it red and flushed; it was pale, sallow-looking, as when he was alive. After he was dead we dressed him, and made him ready for burial. He was left in the cabin until a coffin was made in the fire-room below by the carpenter. When it was ready, we put him into it, took our last look at him, nailed the coffin-lid down, and put the coffin out on the poop-deck. During this time we were making a grave. Tyson, Chester, myself, and several men were hard at work two days digging it out of the solid earth, which was just like flint, with crowbars and pickaxes. We finished it, and on the second day, the 11th, we carried him there and buried him on a flat piece of table-land on Polaris Bay, opposite the ship's winter-quarters. Regular service for the dead was performed by Mr. Bryan, the astronomer, a son of the Rev. Mr. Bryan. The service was read by the light of a lantern held for that purpose. It was dark then—the arctic night. After Captain Hall's death, it appears that there was divided authority, as near as I could understand. I heard that Dr. Bessels had authority, and Buddington went among the men and made very free with them, and of course told them he was captain also. But I always recognized

Captain Buddington as the captain of the ship. There was nobody who questioned his authority as captain of the ship that I know of. During the winter we got along very well—peaceably together. There was nothing of importance occurred that I know of that is worth mentioning· I might think of something if my mind was directed to it. The ship broke adrift after awhile, after we banked her up. We banked her up to keep the frost from penetrating to the interior of her. She broke adrift in a gale of wind, and fortunately she drifted against Providence Iceberg. That saved us from going out into the pack and probably being lost, or driven it is impossible to tell where. We made fast to that berg during this heavy gale and darkness. In a couple of days afterward the young ice formed outside of us. It was several inches thick, and Captain Buddington had that sawed out, and a bed made for the ship a distance from the iceberg—a safe distance, as he thought, for the winter. Shortly afterward, when nicely frozen in, a gale from the southwest came on, and drove the pack against this iceberg, and drove the iceberg in-shore with it, and right up against our vessel—in fact, drove a spur of the berg in under our bows. She lay in that condition all winter, and at low water, at the fall of the tide, this forward part of her would rest on the spur of the berg. It made a cradle for itself in on the spur; and at low water she would keel over, and at high water she would come up again. She was going that way twice in twenty-four hours during the winter; and when the spring came, and the ice began to melt about her bows, the water began to come in in a stream, and we found, then, that her stem was displaced, and a crack at the six-foot mark came from her stem as far down as we could see her—seven or eight feet. She had been wrenched on the berg, and her cut-water slewed to one side, and opened on both sides. There were attempts made to prevent the water coming in, but they did not succeed. Then we made a water-tight compartment, but the water flowed over the bulk-head, and in among her upper works and down through her timbers.

The attempts that were made to relieve us of the water failed, and then we had to put the donkey-pump to work to keep her free. The water came in steadily and constantly. After the ship broke loose, the first time, we certainly could have taken her back to the old floe from which she had broke off. Part of it stood there, and was not more than one hundred yards, but it was my opinion that she was safer where she was, if the iceberg had kept a certain distance from us; but when the iceberg came up to us, I have no doubt that if she had been taken away from there at the time, that she might have been prevented resting on it; but I do not know that there was an effort made to do that. She rested there during the winter. During the winter I never left the ship except to go on shore for provisions, and then came right back again. I had charge of the provisions until Captain Hall died. I did not have charge afterward; I found it would be an unpleasant situation, and I gave charge to Captain Buddington, with the keys, and resigned. I did not have anything to do with provisions, clothing, or anything of the kind. Captain Hall had previously given me charge of all these things. I had a knowledge of accounts, and was familiar with these things, and I suppose it was for that reason that he gave them to me.

When the spring opened, we got all our provisions from the shore, and put them on board the ship again, and we resumed the summer rooms, and put provisions in the lower cabin, and made everything ready for sea and to pass the summer with. We unhoused our ship, took the canvas off and dried it, and put it away. That being done, on the 1st of May, Captain Buddington detailed Mr. Chester and Mr. Tyson to go on a boat-journey. In the mean time Captain Buddington and Dr. Bessels had an understanding. One was to conduct the sledge-journey and the other the boat-journey, but the sledge-journey was left to Dr. Bessels, in fact altogether. He had charge of them, I understood, and could do as he pleased, go when he liked, and organize a party when he liked, and so on. But a boat party was proposed by Captain Buddington, and he said he would take charge of it himself. He did not, however. He detailed Mr. Chester and Mr. Tyson to take command of them, and no sledge-journeys of any consequence were undertaken. On the 3d of June, however, the boat parties were ready, and I believe started, dragging their boats to Cape Lupton, a distance of about four or four and a half miles, to Robeson Channel, where there was open water. They started thence some few days afterward. Chester had the mishap to lose his boat in a few hours after he started. It was sunk with everything on board of her. Tyson did not start for a day or two after, but went ahead as far as sixteen miles up to Newman's Bay, and was there stopped by the ice. Mr. Chester returned to the ship and requested the canvas boat, so that he might try again. His party volunteered to go with him, and he got supplied again and started after a few days. He went up to where Captain Tyson was. The party consisted of Mr. Meyer, Mr. Chester, and Dr. Bessels, Captain Tyson, and four seamen in each boat. At this time there had been no sledge-journey made of any account except when hunting-parties were gotten up. They went out on sledges, but the season for sledging was then over. There was no ice or snow on the shore, and the ice in the channel was broken up, and the snow was soft, so that this rendered the season for sledging over. A sledge-journey should be undertaken early in the spring, in March or April at the furthest. In the mean time a gale of wind came up and broke the ice within a short distance of our ship. When we found it so we commenced sawing, and by sawing and heaving the pieces out for several days we succeeded in freeing our vessel. The heave of the sea coming in from the channel, and from the bay, it broke the ice up, and being previously sawed in several places around the vessel, it broke into different pieces and drifted away, and the ship slid off out of her bed in the berg, the same as if she was going off the ways into the water, and so she got afloat again. That was about the 26th of June. We went to sea that same evening that we broke out, and went into the channel. It was perfectly free from ice for a certain distance. The bay was a mass of water all over. There was scarcely a particle of ice to be seen in the channel. We went up there near some of the capes, pretty near to Newman's Bay, to the south cape of Newman's Bay. There we met a heavy pack of ice, with no chance for the vessel penetrating through it. We fired three heavy shots out of our twelve-pounder howitzer in order to attract the notice of the rest of our party if they chanced to be near. I heard afterward that some of them heard the reports but could not understand what they were, because they had no idea of the ship breaking out at that early period, she was so imbedded and surrounded with the hummocks and broken ice and icebergs. We then came again back to our winter-quarters alongside of the berg. In the mean time two of Chester's party came down. It seems that they had got short of bread. This was towards the latter part of June. We were very poorly manned on the vessel. There was only the captain and myself that knew anything about sailorizing. The rest were Mr. Bryan and two firemen and two landsmen, and a few others, with the cook and the steward. Captain Buddington con-

cluded to retain these men, at least one of them, and sent word to Chester that the ship was broken out and making water freely, and that if
there was a chance to get north we could do it with the ship as well as
with the boats. We finally landed the men with a bag of bread up at
one of the capes at Cape Lupton. They were a good while getting to
their camp with the bread, and we returned to Thank-God Harbor, and
made fast to the berg. We sent Hans before that to tell them, but
Hans brought word back, and the doctor came along with him, and
then we dispatched some men with the bread to Chester. After that
we went out again, but could not get up as far as the boats were, and
came back the third time to Thank-God Harbor and made fast in our
old winter-quarters again. A short time afterward Captain Buddington
sent a note, I think, requesting them—I do not think he sent an order,
he was not firm enough in that respect, I believe—but he sent a request
to have them return. He made known to them the condition of the
ship, and told them that they would be of more use on board the ship
than where they were lying up in the ice. A few days afterward Tyson
and his party came down, and in a few days after that Mr. Chester and
his party came down. We were then all on board the ship again, minus
two boats and the canvas scow that was left up in the channel. We did
nothing particularly after that. We had a good deal to do to save our
vessel. We got aground three or four times, but the ground under us
was soft, and we got her off each time. Finally we found there was no
prospect of doing anything. The season of sledding was over, and the
channel was full of ice. I do not know that there was any consultation
about it, but the first thing I knew we were on the lookout for water,
to go south with, and were under orders to get under way. We slipped
our anchors and did not get either of them, and came down the bay
toward Kennedy's Channel. At this time we met a great deal of obstruction by the ice flowing out of Robinson's Straits into this bay. We
had a good deal of difficulty in getting along. In some places we
would get a lead for a short piece, and then we would be obstructed, and had to bore our way considerably. We could not
force the vessel as much as we could have done on account of our disabled bow. She was a fine vessel, as strong a one as ever I put foot
on. She was well provisioned, well provided for in everything; she
was well supplied in every respect but in regard to coal. She was not
able to carry enough coal for such voyage, owing to the long delays to
which we are so often subjected, and the obstructions to be met with.
Coming through Kennedy's Channel we were beset a few days, but in
no danger at all. We finally got out of it and got into Smith's Straits,
and had a good prospect of getting home by the fall. There appeared
to be a good many leads along the west shore, and a good many running
out into Smith's Straits, but a person with any judgment at all, that
knew anything about Smith's Straits, would never get out into the heavy
pack that is known to exist abreast of Humboldt's Glacier There are
innumerable icebergs there, and a pack of ice the whole year around for
years. By some mishap—I suppose, it was done for the best—the ship
went into a very favorable looking lead out into this heavy pack, and
got beset. There was a great deal of effort made to bring her into shore
again—to the west shore—but it was almost impossible. We bored
and did everything that could be done, but met with no success.
We were finally beset and made fast to a heavy floe, or pieces of
table-ice—a good large piece that was probably several years old. This
was outside of Kennedy's Channel, and up probably in the neighborhood
of Cape Frazer, or up at the head of Smith's Sound. We were in sight

of Cape Andrew Jackson, and could see the west end of the glacier at the time. We drifted then continually. Some days we would drift a good deal, and some days but very little. Occasionally there would be a lead of water for a small space, and we got under way several times. On one occasion we went from one large floe to another and made fast to it. This last time we made secure and fast to a very heavy old floe. We were expecting still to get a lead to the westward. Inshore to the west land, not a great way from it, we saw leads of water that if we could have gotten into them we might possibly have got down south and made our way home. We were unable to do so, however. The young ice commenced to come so rapidly now that we finally found ourselves housed in, seemingly, for the winter. The young ice began to make around the ship. We were able to travel over it and go on to this old floe and dig wells. The wells are formed there by snow melting in the hollow; they are sometimes four and five feet deep, and often there are two or three feet depth of fresh water. We supplied the ship with water from them until we finally broke out. The wells are coated with ice, and we had to break a hole in them every morning and get our supply of fresh water for the ship from them. We continued drifting in this pack, drifting to the east shore considerably. We saw the east shore pretty much, at least until we got pretty near Kane's winter-quarters at Rensselaer Harbor. At that time we thought we would be driven in there with the pack. We were not more than thirty or forty miles from the harbor, and we were in hopes that we would be driven in there and there stopped, so that we might winter in safety and be able to break out in the spring again. The ice, however, took another turn and swept over to the other shore in a contrary direction. We went rapidly then down through Smith's Straits and by Littleton Island. That is in Smith's Straits, between Baffin's Bay and Smith's Sound. We went rapidly out through that, and coming down past Cape Alexander, abreast of Southerland Island. We then could see Northumberland Island, probably fifty miles distant, to the southward of us. This occurred about the 15th of October, and on the night of the 15th, early in the evening, between 7 and 8 o'clock, it commenced to blow a hurricane. Before that time we had a house on the ice, and some provisions and some clothing put into it for safety. We also had all our provisions that were required, in fact, put on deck—some aft and some forward—and five tons of coal. On the night of the 15th of October it commenced to blow, and the ice outside of us, that formed since we got beset, moved away. It left one side of the vessel all water. The ice finally came in from the outside of us; that is, the ice that receded from us came in again and nipped us severely and canted the vessel over considerably. On this occasion there appears to have been a flow of water below, that ran from one portion of the vessel to the other. The engineer ran up and reported that she was badly nipped on the quarter, and that she was stove in, and that the water was rushing in from aft. Captain Tyson got that report, and in my hearing told it to Captain Buddington. Thereupon Captain Buddington ordered the provisions and things that were prepared to be got overboard, to be taken on the ice, and ordered a certain portion of the people out to receive them. I thereupon went aft, where there was a great deal of provisions, and Mr. Bryan was with me, and Mr. Chester, with the other men forward, and some of the men were at the pumps—at the small alleyway pump. It took a few men to relieve each other at that, and then we commenced putting provisions and stuff overboard. We had nearly completed all this, and we were all, in fact, intending to go on the ice

for the purpose of waiting to see what would turn up, at least. We
knew that, even if all the hands were on board the ship, we could not
save her through the winter. We knew that the pumps would freeze
up; and we had not coal enough to keep her going during the winter
season. We felt that we would have to let her sink under us. That
was my opinion, any way; and I think the others entertained the same
opinion. If we had had coal enough we might have saved her. If we
had been able to pump her all winter we could have saved her. When
we had almost everything done, and just waiting for a few moments
just to see what would happen, her stern-hawser snapped and broke.
It pulled out all its fastenings, and then the strain came on the other
one, around the main one, and it snapped also. She swung off, and the
whole strain came on the bow-hawsers; but in some way or other that
parted. Some say that that slipped, or that the ice-anchor drew. Any-
how, we did not get the anchor on board, but we got the hawser. It
was blowing such a terrific gale from the southward that we went like
a shot out of sight, and did not know where we were going at the time. We
wanted to be on the ice, and it appears that some of the men on the ice wanted to be on board the ship. I would have preferred being on the ice, regarding that at the time as the safest place. All our effects were on the ice; all
our clothing. We had not a stitch except what we had on. There was some
bedding, and some clothes the crew left, but that was all. We fortunately
had some provisions, about enough, but not much more than enough, to
last until spring. The bulk of our provisions and clothing, and everything that we had, with about twenty musk-ox skins, were out on the
ice, and we were very sorry we could not be there with them. We
thought the ice was the safer place of the two. I do not think there
was a man on board the Polaris but thought this, and I think the men
on the ice thought so too at the time. They were anxious enough to be
out there; but some people had to stop on board and send the things
out of the vessel. When we drifted out the donkey-engine was not going. The engineer was ordered to get up steam on the little boiler as
fast as he could. He did so by burning everything he could lay his
hands on. The water was still making on us, and coming up near the
furnace. The donkey-engine was out of order; but there was hot water
in the boiler, and we put the boxes in the deck-pumps and poured buckets
of hot water into that and thawed it out. The deck-pump is a powerful pump. It is able to force out a great deal of water. We finally got
it to working, and that actually saved our lives. It could not save the
ship, but it saved our lives. We pumped the water out of the ship, but
it would not run off the deck readily, and came around our legs, and got
solid where we were standing; and we had to shove it back so as to
give the other an opportunity to come out. It was all forced up on
deck in a slushy state. We continued at that until Mr. Schuman reported steam; and never were men better pleased in their lives than
we were to hear that steam was up, thus knowing that we would be able
to keep her free. It was like being rescued from death almost. When
we got up steam we were able to pump her by steam. In the morning,
when day dawned, we found ourselves up Smith's Strait, north of Littleton's Island, and some three miles or three miles and a half from the
shore. There happened to be a "lead" of water inshore from us. The
wind then continued at this time to blow a little, not a gale, but a nice
breeze from the southeast. We commenced to drift down again out
from the pack from where we drifted the night before. We made every
effort to get into this "lead" of water. We could not keep the ship afloat
long. We found there was no use in our trying to save the vessel; and

if she had gone down where she was we might as well have gone down with her. We could not have saved anything, probably not even our lives. We made every effort we could by sailing; and Schuman, every two or three minutes, would use the steam, which he would keep up for two or three minutes, in order to give the vessel a little push ahead. The little boiler was not able to keep steam in her. We finally succeeded in getting into what was formerly known as Life-Boat Cove, where Kane buried his life-boat, going up in 1853. I knew the place the moment we landed there, on account of its vicinity to McGarry Island. In the mean time the first thing that was done was going to the "crow's-nest" and the mast-head. As soon as daylight came, and very often afterward, Chester went up to the mast-head, but said he could see nothing. He saw a black speck on the ice, but he could not tell exactly what it was. It was not moving, however. He thought at first it might be barrels, but we came to the conclusion it was the shade of a piece of berg or hummock. Then Henry Hobby, I think it was, the man who previously was on the lookout very often, went up to the crow's-nest during the day for the purpose of seeing our people if they were in sight, but no vestige of them was to be seen. We finally succeeded in the evening in getting the ship in as near shore as possible. It happened to be high water, and we made her fast to the grounded ice—I mean those heavy floes that rest on the shore, and that sink down and go to the shore at low water, and float at high water, and that are driven out sometimes. With a whale-line that we had on board, and a piece of hawser, we made her fast to the inside of the floating pieces of ice, to the ice that was fast on the shore, and at low water she rested. She took the ground. She was several feet out of the water then at her bows; and we went and examined her, and her stem was completely knocked off. I wondered how she floated so long. She could not have stood long in the condition she was in if she had had anything at all to contend with. Her stem was completely knocked off, and a split as far as eight feet along her bows where the old wrench was. We were only too thankful to get in. We commenced immediately to take down our sails and spars in order to build a house on shore. That was commenced at once. After we got them off we conveyed them ashore. Mr. Chester, the carpenter, and Booth, I think, went to build the house, and I went with the rest of the party and commenced getting everything out of the ship. What provisions were left, and everything that was movable, were brought ashore to where the house was built. We were several days at this work, assisted by the Esquimaux, but we were able to sleep in the house the second night. The Esquimaux came the second day. Two of them I was formerly acquainted with, Myonk that was with Dr. Kane awhile. He was an old acquaintance of Dr. Kane's. I went out to meet them on the ice, and I recognized Myonk. When I spoke to him a few words as well as I could, he recognized me, and I brought him on board and introduced him to Captain Buddington. He stopped with us a few hours, and helped us with their dog-sledges to drag the things across the ice. We had a great deal of difficulty in doing so, and fell through a good many times. He went down to his settlement, and the next day we had five or six sledges up. We went to work, and in a very short time we had the vessel stripped and nearly everything ashore. We then made arrangements for the winter. We covered our house with the sails and got our coal on shore, which was six tons at the most, and what spare wood was about, and covered that and our provisions. We built an outside shed for them to save them from the inclemency of the weather. The Esquimaux staid by us all this time

5 P

until everything was arranged. A few days afterward Captain Buddington made them presents of what materials we had. We had a great many spears, harpoons, and things of that kind, and needles that were left on the ship. The best part of our trading articles were in the large chest that Captain Hall had for trading purposes, and it had been put overboard. It was a great loss to us—in fact it was a loss to the Esquimaux, because it would have been of great benefit to them. Then after that we commenced our winter on shore. We were comparatively comfortable. We had berths all around the sides of the house. We covered the top of it with snow to prevent the frost coming in, and we put a stove inside of it, and a cooking-stove adjoining the outside door, and Dr. Bessels and the scientific gentlemen put up their scientific apparatus and attended to them during the winter, and thus time passed. We collected ice from recent icebergs convenient to the house to melt water for drinking*and cooking purposes. Our coal gave out with the exception of two bags that we kept for blacksmith purposes, that is, to build our boats with. We did not use them all. Some of them were left there in our winter-quarters.

In February we had to resort to the ship; by lamp-light in the first place. We took her spare rudder and sawed that up for fire-wood, and we took her bowsprit out of her, and then the masts, and then we took her house away, first selecting the boards for building the boats; in that manner we provided ourselves with fuel until we came away. Mr. Chester commenced to build his boats. The weather prevented him a great deal. Some days he could work for a few hours, and some days he could not do anything. It was very cold until late in the season, but he contrived to build two admirable scows. They were very well built indeed. They were better than I expected to see. In the mean time Dr. Bessels collected some particular things that he wanted saved. I disremember what was in that box now. In fact I never was acquainted with its contents. There were three boxes left up on the hill in a cairn there. They were left in charge of the Esquimaux. There were presents given to the Esquimaux too. Some of them remained permanently with us, almost all winter, and at the time we left there there were two families there. They were threatened that if they touched the things we left they would be badly dealt with; that they would be punished. They said they would not. After that we were ready. We got our provisions in the boat. We had them previously made up and provided while Chester was building the boats. I was superintending, under Captain Buddington's directions, the putting up of the stores to be carried in the boats, and a certain amount of clothing. We had not a great deal. There was only a certain amount allowed. The 1st of June was our time for leaving. There was then a gale of wind. The next day was Sunday. The gale continued on that day, but on Monday morning, I think it was, we started. We had open water round Cape Alexander and down to Etah Wetany. We passed through and tried to get further down toward Northumberland Island, but had to come back there again. We wanted to follow the shore but found pack-ice in there. We came back and remained that night at this Esquimaux settlement. The next day we had to go out in the bay, outside of the ice, through a lead, through an innumerable row of icebergs, sailing in and out through them to Northumberland Island. The distance between where we were and Northumberland Island was more than thirty miles, but we made it rowing and sailing.

When we had a fair wind we could make a good way sailing, but when we had a head wind, or a calm, we had to row; we made North-umberland Island late that night—somewhere about midnight. That was a good day's journey. It was Hakluyt Island where we rested. We remained there for a couple of days on account of bad weather, and then went over to Northumberland Island. There was a good deal of ice in the vicinity, and we made two or three attempts to leave it, but could not, and we rested on Northumberland Island in two or three different places. We finally started across to the main-land, toward Cape Parry, but we got beset and stuck in the ice, and drifted a part of that day and all that night in the pack, and in great danger of being lost. We were drifting out, heading our boats on a small piece of ice that drifted out into Baffin's Bay. But we finally succeeded in carrying our provisions from one piece to another, and our boats afterward to the same, and so on, until finally we got to a lead of water, and succeeded in getting to the place we left the day before, in the same spot. We again started after a short rest to the main-land, and succeeded in getting past Cape Parry. We went along the shore then toward Sanderson's Island, and went on to Wolstenholm's Island. We remained there a short time to bivouac, and finally came past Cape Dudley Digges and Cape York. We got south of Cape York, having had in the mean time a great deal of difficulty with floe-ice. We succeeded in getting some twenty-five or thirty miles south-east of Cape York, in Melville Bay, when we were obstructed again by the heavy pack. We were alongside the fast ice, but the broken floe-ice was close against it.

About this time one of our boats got injured, but not very badly, and we repaired it. While here contemplating what we would do next we espied a whaler some ten or twelve miles distant; we sent two of our people to communicate with them, and tell them of our situation and who we were. Before they reached the vessel, there were a number of men dispatched from the vessel coming toward us, some twelve or fifteen. On meeting with our people two of them went back to tell the captain of the vessel who we were; the rest came on. That evening we started, with what effects we had or could carry, some small things needful, and, assisted by the crew, we got on board the Scotch whaler Ravenscraig, of Kirkcaldy. That is the name of the place in Scotland where she belongs, though she sailed from Dundee. Captain Allen received us very kindly and attended to our comforts, and was assisted by the surgeon of the ship, whose name I cannot now recall. We were very thankful for the kindness shown us, and for falling in with these men, because there was difficult work before us. It was very doubtful whether we should ever reach Upernavik. Our boats could not stand a nip; they were very slightly built, and our provisions we were afraid would give out before we got there. We had only just commenced our journey, and we had the most difficult part of it to do yet. After we got on board the whaler, everything was comfortable and pleasant. We felt greatly relieved because we thought that we were safe. After the ship got out of the ice and over on the west shore at Lancaster Sound, on account of the small quantity of provisions that he had for such a number, the captain distributed us to different vessels. He sent some to the Arctic, some he kept on board the Ravenscraig, and some were put on board the Intrepid. Afterward, when the Arctic got filled with oil and was returning home, we transferred ourselves from the Ravenscraig to her, went to Dundee, and came home from Dundee to the United States by way of England, reaching here in the City of Antwerp. Three of our number were left on board the In-

trepid—Mr. Bryan, Mr. Booth, and Mr. Mauch. It is nearly time they
arrived now. I expect them almost every day. It is nearly time, because there was another ship about coming home very soon, or about
the same time that we did. All that she wanted was one whale, and
probably she may come without it. There was also another ship leaking
badly, and the captain said he had reason to come home, and probably
started about the same time that we did, and if they fell in with either
of these ships they are nearly home now. They cannot be much longer
away, because the whaling-season is past, and they must leave whether
they have whales enough or not. They would not be likely to winter
up there; they are not prepared for winter; they must leave before
winter sets in.

I have given you an outline of our proceedings and adventures as
far as I recollect. I kept no journal. Perhaps once in a month I would
note down in a pocket-book one little incident or another, but that
was all. It was a little book, and I did not care anything about it. I
understood by report that there was something similar to it picked up
on the ice. This little book, which was a little book with a leather
cover not unlike a bank-book, was found on the ice after we had started
from the party on the floe, and that is similar to the one I had.

(The printed volume of testimony containing statements of the parties on board the Polaris being shown to Mr. Morton, and he being
asked if the extracts which purported to have been written by himself
were his and were correct, he said they were, but that he did not expect
them to be seen by anybody when he made them, having just written
them down for his own use.)

Question. Have you looked at this chart of Mr. Myer's?
Answer. Not very minutely.
Question. Look at this chart and state whether it is generally accu-
rate.
Answer. I think it is. As a matter of criticism, I should say that
Newman's Bay did not appear to me from the ship on the day we passed
it to be as wide as it is laid down on the chart, and it seems to me that
Polaris Bay is not as much of a bay on the chart as it really is.
Question. You are the very man of Dr. Kane's expedition who was
at Cape Constitution?
Answer. Yes, sir. Hans Christian and I were there for Dr. Kane.
We were both on board the Polaris on this expedition. We partially
recognized Cape Constitution as we went up, but we were certain of it
when we came down.
Question. Is it correctly laid down on this map?
Answer. I should say it is. It appears to me, however, that Cape
Andrew Jackson is nearer to Cape Constitution than is represented on
he map.
Question. Do you think Cape Constitution is put in the right latitude
on the map?
Answer. I dare say it is. When I went there for Dr. Kane I considered it by dead reckoning to be some forty miles up the channel from
where I started. I made no survey this time. I merely looked on.
Cape Andrew Jackson appears to be farther down than I supposed it
to be, but coming down we recognized Cape Constitution to be the
same place that we supposed to be it going up, and we saw the exact
spot where Hans and myself killed two bears when we were on Dr.
Kane's expedition. I do not know whether its latitude is accurately
laid down on this chart or not, but the same place that we supposed to be
Cape Constitution as we went up is the Cape Constitution I visited with

Haus, and it bears the same relation to the rest of the shore-line as is shown on this chart. I could not have been accurate when I was there first in regard to the latitude. I had only received a lesson in taking observations from Dr. Kane and Mr. Songstag before I started, and I was only an amateur; by no means a proficient. But we recognized the place this time, particularly as we came down, as the same place at which Hans and I were then, and although I cannot be certain as to the exact accuracy of the chart, it bears in reality the same relation, so far as I can see, to the coast-line above it as laid down here.

Question. Do you remember Kane's chart?

Answer. I have only seen it once, and it was a long time ago, when it was newly made. It has been revised since I saw it.

Question. Do you remember on that Kane's chart, where you put down Cape Constitution, land breaks right off to the west?

Answer. That was only a supposition that it broke right off, because we did not see it. We did not get far enough in front of the Cape to see around, and it was only a supposition that it broke off.

Question. Do you remember that what is now laid down in this chart as land on this shore above Cape Constitution was laid down as water on Kane's chart?

Answer. Yes, sir; it was supposed to be water. The land was not seen at that time. In sailing up this time we discovered that instead of water above Cape Constitution, there was land more than thirty miles above up to the opening now called the Southern Fjiord, or above that, on the shores of Polaris Bay and Robeson Channel.

Question. After Captain Hall died, did you hear anybody express himself as relieved by his death?

Answer. I did not; but I thought some people were not very sorry. I did not indeed hear any such expressions. I never heard Captain Buddington say that he was relieved, but I heard within the last few days that he did say so; I did not hear it myself. The discipline of the ship was good during Captain Hall's life-time. He was a very kind man, but strict. There was nothing tyrannical about him. Still everybody appeared to dread him and respect him. That was my feeling toward him. I did not dread him, but respected him very much; I was an old man-o'-war's-man, and discipline was familiar to me; after he died the discipline was loose, and every person did almost as they pleased. Still, I saw no bad actions, or acts, committed. Captain Hall and Captain Buddington during Captain Hall's life-time occasionally had a few words. Still, there was a good feeling between them. They appeared to be indebted to each other for favors and kindnesses. Still, Captain Hall had a few words with him—I suppose in the line of discipline, and things of that sort, on two or three occasions.

Question. Did you ever hear Captain Buddington depreciate Captain Hall among the crew?

Answer. No, sir. I have heard him mutter to himself inarticulately several words. I did not want to listen to what he was saying, but I knew they were a little disrespectful to Captain Hall.

Question. Did you ever hear him talk among the men disrespectfully of Captain Hall?

Answer. I did not; but I understood he did. He has used a good many careless expressions that I did not take notice of. He was very foolish in a great many of his expressions; and I did not think the man meant what he said half the time. I saw him under the influence of liquor a couple of times; but could not swear that I saw him incapable of doing his duty. I know that he was boozy and intoxicated, but still a man can do a good deal when he is even that way. I never saw him lying down dead drunk. I heard that he had some difficulty with the doctor about the alcohol. I heard a slight altercation between them. I was in the upper cabin, but this happened, I understand, in the lower cabin. I was told it both by the doctor and by Buddington himself. I was told that the captain was tippling on alcohol; and the doctor proposed to watch him, and hid himself down in the lower cabin, in Hans's quarters, and he lay there in ambush until the captain came down, when he had a little bottle secreted there, and as he came down there and took his nip, the doctor sprung out upon him and wanted to snatch it from him, and Buddington got hold of the doctor. I do not know whether one of them fell or not. But such a thing happened between them, I understood, and I think I heard the scuffle going on; but I was not present at the time. It was told by both of them afterward.

That night after we left winter-quarters and were coming home, when we left the west shore and got into the middle of the channel and were beset, I cannot say that he was drunk, but he had been drinking. I saw him able to give orders and work on deck, but I should think that he had been taking something. I know Captain Hall kept a journal, but I do not know what became of it. I saw it after Captain Hall's death, but I have not seen it for a long time. It was kept in his desk. It was in a large book like this. (Referring to one of the regular printed Navy journals.) After his death I saw some people reading it. I think I saw Captain Buddington and others read it. I disremember who else; perhaps Mr. Chester; I am not sure. I do not know whether any others read it or not. I saw them reading it in the cabin. This was not very long. It was, I understood, put away among his other papers. They were put, I understood, in a tin box. He had a writing-desk and a tin box with a lock and key on, and Captain Buddington put his papers in it and kept the keys of it. That box was on the ship until the evening we broke away; after that I do not know what became of it. I did not see it. It was not in the ship after we went ashore. It was in the cabin aft, I suppose, and Mr. Bryan and I were aft at the time the ship went adrift putting the things off, but I do not remember of putting that box over. I am sure I did not. We put pemmican and other boxes of meat there, and clothing and bedding, &c., and a heavy bag of ammunition, with powder, shot, and every other thing in it, I had prepared and laid in the wheel-house, but I am sure I did not handle that box to put it out. Somebody else may have done so, but I did not. I did not see it afterward, never. I would know it if I had seen it. I have never seen it since that time. It was a japanned box, with a padlock and key on it. I did not see anything of any of Captain Hall's papers after that. I saw his papers frequently in Captain Buddington's hands, just merely to replace them in his box, but I have not seen any of Captain Hall's papers since that night. I do not know that any part of the journal was burned. There was something burned, but I do not know that it was any of the journal. This was a few days or probably a day before Captain Hall died. I understood that it was burned over a candle in the cabin by Captain Hall in the presence of Captain Buddington. I heard this from Mr. Buddington himself; that Captain Hall had written a letter against Captain Buddington, and that he said that as long as he intended to destroy it, it was not worth while that he should read it. He said that it would only leave something bad in his memory, and he would destroy it. It was burned. I saw the burnt parts of it on the table that the candlestick was on that night.

Question. Have you any way of accounting for the fact that you did not see the men on the ice the next morning after you went adrift?

Answer. I have no idea. They must have drifted. There appears to be no doubt about the fact that they saw us. They all saw us, it seems. And there is no doubt about the fact that we did not see them; I am not able to give the reason, unless they were under the lee of Littleton Island; they may have been one side of it and we the other, and we could not see them on account of that. They were low down. Mr. Chester and other men went frequently up to the crow's-nest, but did not see them. Captain Tyson could not have known where he was, or he would not say we were at Northumberland Island. We were not at Northumberland Island. We were sixty miles from it. We didn't go an inch in the direction in which the party were. We went directly in to the southeast— directly ashore as near as we could and toward Littleton Island, not more than two miles above it. The only reason I can give that we did not see them was because they were in toward the east shore under the lee of Littleton Island, and we were to the north of it and the island was between us. It is a high island. They might see us, because we were a larger object, but we could not see them. They might after we got in near the shore have seen us in the space between Littleton Island and the mainland. There are two or three miles of channel there. They say they saw us twice as we came down, and afterward when we got into this open channel, but we did not come down. We headed in to shore directly as near as we could. We might have drifted down a little. We did not head down, and tried to avoid going down. We knew that there was no help for us, if we drifted down or did not get this lead into the shore. We could not do anything more than we did do to get the vessel ashore.

Question. If you had seen them you could not have gotten to them?

Answer. We could not indeed.

Question. How much of the time were these men at the mast-head?

Answer. They did not stop very long at a time. They scanned the horizon, I suppose, for ten or twelve minutes at a time.

Question. Were they there half the time in all?

Answer. No, sir.

Question. How much of the time were the men at the mast-head altogether?

Answer. Probably an hour altogether. They would go up and take a scan around, then come down to their work. We were very busy on board of the ship, and all hands had to be to save her. I do not think the men were up more than an hour during the day. We went to work at once to take care of the ship and to get her inshore. When we left Life-Boat Cove we left nothing there worthy of note—a lot of *débris*, old books, &c. They might have been valuable if they were taken care of. We had not the means of transporting them home. The original log was copied into a smaller book, and the copy was brought home, and I think that the original was put into the box, and left in the cairn on the hill up there. There were also some boxes of specimens left there, and instruments.

Question. When Captain Hall had his relapse, what were the symptoms; did he vomit in the first case?

Answer. I think not; not while I was in his presence.

Question. In what way did the renewed illness manifest itself?

Answer. He merely appeared to be like a person who is incapable of using his limbs—helpless almost.

Question. Did that seem to come on suddenly?

Answer. It appears it came very suddenly. He was up to-day, and down to-morrow. I do not know whether it came with a stroke. I was not with him at the time. Mr. Chester was with him. It appears that he took something—medicine of some kind; but whether that was the cause of it I do not know. Of course it was not really the cause.

Question. What induced him to take medicine; because he felt an attack coming on?

Answer. I don't know. I was not there. After the second attack he appeared to be numb. In fact, there appeared to be a general numbness or debility all over him, as far as I know. I did not notice one leg or arm more than another. I paid some attention to the natural history of the expedition. I pulled a great many little flowers, mosses, and picked up a good many stones, and I put away some of them, but I lost them all. I saw some of the drift-wood. It was picked up in Robeson Channel, and brought down to the vessel. I saw it on the poop. I believe it was found on the shores of the main channel of Robeson Channel above Polaris Bay. I only saw what was brought in. Two large pieces of sled and pieces of wood of a pretty good thickness, for wood that grows in that country. I would suppose it did not grow in that country. Upon the question as to whether the tide came from the south or the north, I heard some of them say it came in from the north, but my opinion is it came in from the south; but there was a continued drift from the north. I noticed that. This drift was generally to the southward, whether the wind was blowing or not, in the center of the main channel. I might have seen ice going up for a short time, but it was always sure to come farther south, and generally turned to the south altogether. I saw a good many musk-oxen. I did not see any the second winter, not even in Kane's winter-quarters, which is higher than Life-Boat Cove. He saw the skeletons, but no live ones. But around Newman's Bay there were a good many herds of them. They feed on grass that grows there, and willow that grows there in the summer season. They paw it out in the winter season from under the snow and eat it. I did not see any wolves, but I saw white foxes, hares, and a great many ptarmigans. We could go and catch lemings. I brought in five myself from the shore and had them all aboard for some time. They are innocent little things; they would run away from you, but you could go and catch them without much trouble. They would go under the rocks. They could not burrow down very far. They have hills in the snow in the winter, and have nests made there like a bird. Birds flew there that looked very much like hawks; dark brown in color and as big as chicken-hawks. I think we also saw a good number of ravens at Life-Cove Bay. I am not sure; I do not know whether we saw any at Polaris Bay or not. We saw little brant geese. They were small, not bigger than a domestic duck, but they looked bigger. They had nests there. I do not know that anybody saw their nests, but we saw their young when they came down. We did not look much for them, but we saw a good many of them. They fed along the water's edge, probably on shrimp, and I think on the grass. I did not see any fish, but I saw a good many shrimp, if you call them fish. I think we saw jelly-fish, but I am not sure. There are a great many seals there, a couple of kinds at least. I am not well acquainted with them, however. They feed on shrimp. We saw some northern lights, but not so bright as I saw them down at Rensselaer Harbor formerly. They did not appear to be so bright up there. During the winter we had some very calm and quiet weather; we could see across to the west land from the top of the hills during such weather. During the winter we occasionally saw a great deal of open water. At Life-Boat Cove the last season the temperature was not as low by any means as Kane had it in Rensselaer Harbor. I do not know certainly whether it was much lower than it was in Polaris Bay, but I have an idea that the climate is milder at Polaris Bay than at Kane's winter-quarters. There is less snow at Polaris Bay. The land is entirely bare of snow in the summer season. Down to the south and west there is more snow. It is only at prominent points and headlands where the snow goes off, and all the rest is solid snow. The climate was really milder higher up—in the eighties, than down in the seventies, and there was more vegetation; there must be more vegetation, or musk-oxen could not live. I saw places where they could not exist down a great deal farther south than that. The ice was too thick, and there was no vegetation or anything under it, while at Polaris Bay you would find grass in patches as high as your ankles. There was good feed, and there were young willows that grew up there to a considerable height, some more than a foot above the ground. Down lower, I never saw any more than a couple of inches above the ground; they would spread out over the surface and die away. I think, altogether, it is milder farther north, and there is more vegetation than there is a great deal lower down. Our sensations of cold at Polaris Bay were certainly no greater than at Kane's winter-quarters. We were obliged to muffle up very warmly when we went out during the winter, particularly if it was blowing. If it was blowing at 10 below zero it could not be stood as well as when it was 50 degrees below zero and not blowing. To take any exercise in calm weather, when it was 40 degrees below zero, you would not be able to muffle up much; still, at the same time, your ears, and nose, and fingers, and flesh on your cheek-bones were liable to be frozen. You could not feel it yourself, perhaps, but it would be perceptible in the color of the skin. It would become white, and you would have to get your blood in circulation by rubbing the part affected.

The examination of witnesses having been concluded, the commission adjourned until to-morrow morning at 11 a. m.

WASHINGTON, D. C., *October* 16, 1873.

By invitation, Surgeon-General Barnes, of the United States Army, and Surgeon-General Beale, of the United States Navy, were present to listen to such portion of the statement of Dr. Bessels (who, it was expected, would be next called) as related to the sickness and death of Captain Hall.

Examination of Dr. Emil Bessels.

I was born at Heidelberg, in 1844; graduated at Heidelberg; joined the Polaris expedition at Brooklyn as chief of the scientific department; left New York with that expedition on the 29th of June; next day at noon arrived at New London; left New London on the 3d; on the 11th landed at Saint John's; left Saint John's again on the 19th, making our way for the coast of West Greenland; we arrived at Fiskernaes on the 27th of July to look for one of the natives, Hans Christian, who had accompanied the expedition of Kane, and had been taken back by Hays. We left Fiskernaes after two days, and in going to Holsteinberg we encountered a gale. We arrived at Holsteinberg on the 31st of July. We there met the Swedish expedition, under the command of Captain Van Otter, and obtained some very valuable information in regard to the state of the ice at the north and Upernavik. We remained at Holsteinberg until the 3d of August, when we left and shaped our course to Disco, where we arrived on the 4th of August, at 3 in the afternoon. Finding that the inspector was absent, Captain Hall dispatched a boat, under the command of Mr. Chester, to look for him at Jacob's Haven and Rittenbenk. On the 10th of the same month the steamer Congress arrived. We landed one part of our stores and took the rest of them on board—as much as we could carry.

During our stay at Disco there was a little difference between Captain Hall and Mr. Meyer, and then between Captain Hall and myself. Some kind friends wanted to make out that we had a mutiny on board of the ship. But the whole amount of it was that Captain Hall wanted Mr. Meyer to write his journal, and Meyer did not want to do it. Captain Hall intended to discharge him, and spoke to me about it. I told him that I did not think Mr. Bryan and myself would be able to perform the whole of the work to be done on an expedition like that. I told him I preferred to go on shore myself if Mr. Meyer was dismissed. I saw that we would not be able to do the work. Finally Mr. Meyer agreed to conform to the orders and instructions of Captain Hall, and the matter was settled. Happily, I am able to produce to you the original copy of the original instructions belonging to Captain Hall. I found it when the vessel broke adrift, and here you will find a statement on this page in Captain Hall's own handwriting. I think it explains the matter. (Dr. Bessels produces copy of the original instructions of Captain Hall, containing a memorandum in Captain Hall's own handwriting, and signed by Mr. Meyer, on the 16th of August, 1871, it being a memorandum made at the time of the arrangement of the difficulty. It is written on the sixth page of the copy of the original instructions belonging to Captain Hall, and marked with his name in his handwriting. It is as follows: "As a member of the United States naval north polar expedition, I do hereby solemnly promise and agree to conform to all the orders and instructions as herein set forth by the Secretary of the United States Navy to the commander. Signed, Frederick Meyer, observer, United States Army. God Haven, Greenland, August 16, 1871." This memorandum is written opposite the following clause in the instructions, which is underlined in pencil by Captain Hall: "All persons attached to the expedition are under your command, and shall, under every circumstance and condition, be subject to the rules, regulations, and laws governing the discipline of the Navy, to be modified, but not increased by you, as the circumstances may in your judgment require." Paper is marked by Secretary, No. 1, E. B.)

After having taken some dogs on board, we left Disco on the afternoon of the 17th. We arrived at Upernavik on the 18th, staid there for three days, and dispatched a boat to Proven to get Hans, Kane's Esquimaux, and on the 21st, at 8 p. m., left the settlement, Governor Elburg on board, who proposed to accompany us to Tessnisak to procure some dogs and skins. We stopped at the island Kingituk on our way. We took twelve dogs on board, and arrived at Tessnisak early on the morning of the 22d. We staid there for two days, and left on the afternoon of the 24th.

From this time Mr. Bryan, Mr. Meyer, and myself kept a log. We had two patent logs overboard, one a-starboard, and one on the port side of the ship, and we noted all the distances and courses run. That is the original of our course from Disco up to the highest point north, up to 82° 16′, and our drift back to 81° 30′ to Thank God Harbor. These leaves which I have in my hand were taken by me out of the original log-book and put together, in order that they might be in a more convenient shape to bring home in the boats, because we could not undertake to bring the whole large log with us.

These contain the original entries made at the time in the log-book of the courses and distances and other remarks made at the time. This log was kept by Mr. Bryan, Mr. Meyer, and myself. (The paper marked by the Secretary No. 2, E. B.) Besides that I have the different courses reduced and corrected. I lost one part, but I kept the other part. Here is the one part I saved, of the reduced courses and distances. This covers the portion from July 3, leaving New London and New York, up to the 26th of July. Here are more documents referring to the same thing, showing some of the courses of the Polaris. These are taken from the ship's log, because we only kept a log after leaving Tessnisak and at Smith's Sound. This little book (producing a little book resembling a bank-book, with leather cover) is our rough log. (The paper taken from Mr. Chester's log is marked by the Secretary No. 3 E. B. The rough log is marked No. 4 E. B.) We left Tessnisak on the afternoon of the 24th of August, and passed Cape York at 7.45 in the evening. There we met a little ice. On the 26th, at 4.30 p. m., we experienced for the first time a northerly set, indicated by the drift of ice moving rapidly to the south. At 7.30 we passed Cape Parry, bearing northeast by east, distance about twenty miles. At 8 o'clock we passed many bergs aground, abreast Cape Parry, imitating outline of the coast, seeming to indicate a shoal lying off the coast. The same range of bergs we also saw during our retreat in the boats in June, 1872. At 10 in the evening we found ourselves surrounded by broken ice. We had to steer very irregularly to avoid collisions, always keeping the land on the starboard side. Latitude at noon, by dead reckoning, 76° 12′; longitude, by dead reckoning, 69° 37′. By observation we made it 75° 56′; longitude, 69° 26′ 30″. At 1.8 o'clock in the afternoon we passed Conicle Rock, fifteen miles distant; at 2 o'clock, Cape Dudley Digges, about twelve miles distant; at 6 o'clock we saw a great number of walruses, and tried to kill some, but we did not succeed; at 8 in the evening we passed the mouth of Granville Bay, and an hour later we were compelled to take the logs in, being surrounded by broken ice. We put them overboard again at 9.30, but had to take them back 20 minutes after that. At 11.10 we passed Fitz Clarence Rock. At 4.30, on the 27th, we sighted Cape Isabella and Cape Alexander, at the entrance of Smith's Sound. At 5.15 we passed Hakluyt Island. Five minutes later we were stopped by ice. Latitude at noon observed 77° 51′, longitude at 3.51 p. m. 73° 5′. At 3 in the afternoon we entered Smith's Sound and passed Cape Alexander. At 4.37 we passed Port Foulke, at winter-quarters of Hayes, and at 5, Littleton Island; 6.50 we passed Cairn Point, and at 8 we found ourselves abreast of Rennselaer Harbor.

Now we began to shape our course to the west. What seems remarkable there is, that instead of finding the western shore blocked by ice, we really found there open water. We shaped our course to the west, not because we were met by ice, but because it was of the utmost consequence to follow the coast-line, and the east coast trended a good deal to the eastward, but we would not make as much north by following it. Consequently we took to the west course, and got along in a very short time. But what I want to say is that every current moving in the direction from north to south will be deflected to the westward on account of the rotation of the earth, and consequently it will deposit its ice, or any foreign matter that it carries, to the westward. In point of theory we would expect to find that, but in reality we found it to be different; and we found this to be the case at every island or continent in the arctic regions. So, for instance, the west coast of Spitzbergen has been explored thoroughly. The east coast is hardly known on account of the ice. We find the same thing in Nova Zembla and on the east coast of Greenland. I do not know how to account for it. On our expedition we found in going over from the west coast open water where we might have expected ice to be deposited; we had the ice to the starboard side going out.

Question. How is it in Greenland, farther south? Is the ice on the east coast or west coast?

Answer. The ice is on the east coast of Greenland, and has accumulated there. The east coast of Greenland has been visited but very seldom.

Dr. Bessels, (resuming:)

We made over to the west coast on the 28th at 3.30 a. m., and found it clear of ice over there, at Cape Hawks, on the port beam, distant about fifteen miles. There we had to take our logs in. At 9 o'clock we passed Cape Wilkes; at 12.30 we reached Cape Shaw; at 2.30 Cape McClintock. Cape McClintock is a north cape of Scorseby Bay. At 3.45 we reached Cape Lawrence. We found that the east coast of Grinnell's Land is entirely different from what has been given by either Kane or Hayes. I have plotted all the different surveys from the year 1616 to 1865, made by Belot and Baffin in 1616, by John Ross in 1818, by Inglefield in 1852, by Kane in 1853, and by Hayes in 1865. I have reduced them all to the same scale, marking the different surveys by different colors, so that you can see the difference at a glance. (Plan produced.)

Dr. Bessels, (resuming:) At 12.20 we passed close to an island on the starboard side, passing between the island and the land. It had not been laid down by Hayes; but seems to lie in his very track. If his track has been put down right on his chart, he ought to have passed directly over that island. The island lies in latitude . We are still in Smith's Sound. At 2.20 on the 29th we had to stop to repair our engine. We took the logs in and started again at 3 o'clock. At 8.12 in the morning we had to stop on account of dense fog, and at 9.13 the reading of our log showed one hundred and nine miles. Observed latitude at noon, 81° 20′, longitude 64° 34′. As I have stated, we found ourselves at noon in latitude 81° 20′, having passed through Kennedy's Channel. At 4 in the afternoon we met some bergs and broken ice. We sighted Cape Constitution going up as we passed it, but it was not very clearly defined. It was rather hazy at the time, but we could see the land lying above it. At 6.08 we stopped, and started again at 7.18. At 9 o'clock we passed a mass of loosely packed drifting ice. We could see the land on both sides. We have always been able to see the land on both sides all the way up whenever it has been clear after we passed through Kennedy's Channel, the channel being at the widest part about thirty-three miles wide. The next day, the 30th, we found it very foggy, and we made our way through drifting ice, and had to take our log in at 6 in the morning, and put it over again at 9.13. At 9.35 we were compelled to stop, and we reached the third day our highest latitude with 82° 16′, the highest latitude ever reached by any ship. We were in a channel at that time, and some time before that some of the officers thought we were in the bay. When I came on deck in the morning, about 6 o'clock, Captain Buddington showed me a dark cloud, hanging quite low over the horizon, at a pretty good distance to the north, ahead of us. Sometimes, when the fog cleared away, you could get some glimpse of land, and this land is the northernmost land we saw. I placed it in latitude 84° 40′.

It is on none of the charts that have been published, but the land exists in reality. The land runs northeast and southwest. There is a high plateau with deep cliffs. I think we ought to name it Grant's Land. There is no doubt about the existence of land there. A few of us only have seen it, but Captain Hall, in his dispatch to the Secretary of the Navy, on page 15 of the previously printed report, says, "There is appearance of land farther north, and extending more easterly than what I have just noted, but a peculiar, dark, nebulous cloud, that constantly hangs over what seems to be land, prevents my making a full determination."

We made our highest latitude at 82° 16' on 30th of August. In arriving at that latitude we had to construct our course back. It was rather difficult sometimes to do it, but then I think that it will be found to be quite reliable, because we were able to take the mean of two patent logs, and if we did not construct it back, if we took it from our set-point of observation, were deducting the current, it really took us to 82° 29', if we took it by dead reckoning, from that point. But in constructing it back and allowing for current, Mr. Meyer reduced it to 82° 16'. We had no deviation of the compass, and we had no proper observations for variation. So it was rather difficult. We had to take our variations from a chart made up by Mr. Schott for the expedition. I have brought back the original. (Original produced.)

After we reached the highest point we had to make fast to an ice-floe not being able to penetrate any farther. We had a consultation on deck among the officers of the ship, Mr. Chester, Mr. Morton, Captain Tyson, Captain Buddington, and myself. Messrs. Chester, Tyson, and Morton suggested going ahead. I did the same, only remarking at the same time that if we were not able to make any more northing we were to strike the west coast, because we had a fine base of land to proceed on. Captain Buddington said that he did not see any chance to go in farther, and so we did not attempt it. Captain Hall was very anxious to go north.

Question. Was there any opening to the north at that time?

Answer. I had not been at the mast-head. Tyson was there and one of the men, and they both reported that they saw plenty of open water, intersected by drifting ice. I was only on deck, and you know that from there your radius of sight is very limited. It amounted to about seven miles from the deck of the Polaris. We could not see open water from the deck. The ice was intersected by water-leads. We tied up to the ice and drifted back. Captain Hall had before that attempted to land at Repulse Harbor; that is a harbor situated on the north coast of Greenland, but finding the tide running very strong, he came back again. He attempted to put up winter-quarters there, but it did not seem to be very well adapted, being open to the north and subject to the prevailing winds, and consequently would be subject to the drift of very heavy ice.

On August 30 there was fog during the whole time. The rigging of the ship was coated with ice. You could see land on both sides, and could see it plainly. In the afternoon we had a heavy snow-fall, which was very likely produced by being in the vicinity of a heavy pack of ice. At 7.15 in the evening it cleared off, and Captain Hall, with Mr. Tyson, landed again at the same place—Repulse Harbor—but could not get in. It was filled with ice. At 11.30 in the evening we were compelled to make fast again. Ice was moving fast under the influence of the flood-current. August 31, 6 o'clock in the morning, we left the ice-floe; it grew foggy, and we had to tie up again at 7.50. On the 1st of Septem-

ber, at 9.25 in the morning, we tried to push on. We pushed to the
eastward, but about thirty-five minutes after we had to tie up. During
the night there was heavy ice made. Finally, we drifted down. We
could not find any harbor along the whole coast, except, perhaps, at
Newman's Bay, or in that inlet called on the chart Southern Fjiord.
We had to make fast on the lee of an iceberg, called by Captain Hall
Providence Berg, in Polaris Bay. He called that Thank God Harbor.
We went into winter-quarters there. It consists of an iceberg. There
is a slight indentation in the coast, but it is very slight. You would
hardly see it on a map with an ordinary scale. The berg and some floe-
ice formed us a sort of breakwater. We were swept down by the ice.
I do not think that our drift was entirely due to the current. We had
pretty strong northeasterly and northwesterly winds, and the mean
strength of the current amounted on an average from 0.4 to 0,6 of a
mile per hour.

On the 4th of September, midnight, we arrived at Thank God Harbor.
During the next day we were employed in preparing for winter-quarters. The ship was unloaded, and the provisions landed on shore. We
had an observatory set up on the shore at an elevation of 34 feet above
mean sea-level. The Esquimaux were sent out to hunt and found traces
of musk-oxen—animals found for the first time in West Greenland alive.
Kane found several skeletons impregnated by carbonate of lime, but it
is very likely that those animals had existed there a long time before.
Musk-oxen have been discovered in East Greenland lately. The Esquimaux told us that on the east coast of Grinnell Land on the other side of
the channel there are plenty of natives and more musk-oxen. They
hunt them with the bow and arrow.

As already stated, we erected an observatory, and on the 18th Mr.
Chester, the Esquimaux Joe, Hans, and myself were sent out on a sledge-
journey to see whether there was a practical route northward, if during
the spring the ice should not be in good condition to travel on. Besides
that, we went to hunt musk-oxen. We came back again on the 24th,
having found a plain about thirty miles long extending to the northward, and having killed one musk-ox. On the 10th of October Captain
Hall, in company with Mr. Chester and the two Esquimaux, left his
sledge-journey and went up to Newman's Bay, and returned on the 24th
of October. After he came back he was taken sick. He started on the
10th of October and came back the 24th. I was at the observatory at
the time he returned. I had fixed the observatory, and got the instrument ready to take our observations. Up to that time meteorological
observations had been taken every three hours. From the end of
October hourly series began. We noted hourly the height of the barometer, the temperature of the air, the moisture of the atmosphere, direction and force of the wind, and the amount and kind of clouds, with
their respective directions, state of the weather, &c. Besides that, astronomical observations were kept up to determine reliable meridian.

As I say, I was at the observatory when I heard the sledges approach-
ing, and went out to meet Captain Hall and his party. He shook hands
with me, and I accompanied him about half-way to the ship; then I returned to the observatory. After some time Mr. Meyer came over to call
me, stating that Captain Hall was taken sick, and was in bed. That
was about an hour and a half after he had arrived. When I went out
to meet him I had some conversation with him. He told me that he
had had very low temperature and could not make any headway. He
expected to go a great deal farther, but was compelled to return on account of the configuration of the land. The land he found to be mount-

ainous and barely covered with snow, and so he could not make any
northing, and he was compelled to come back. He did not say anything
at the time how he was, but afterward said he had not felt very well
for two or three days.

As I said before, after I saw Captain Hall I went back to the observatory, and in about an hour and a half Mr. Meyer came over to call me,
stating that Captain Hall was sick. I went over to see him. I found
him in his bed. It was rather warm in the cabin, and the first thing I
did was to open the door before I spoke to him. He told me he had
been vomiting, and that he felt pain in his stomach and weakness in his
legs. While I was speaking to him he all at once became comatose. I
tried to raise him up, but it was of little use. His pulse was irregular—
from 60 to 80. Sometimes it was full, and sometimes it was weak, and
he remained in this comatose condition for twenty-five minutes without
showing signs of any convulsions. While he was in this comatose state I
applied a mustard poultice to his legs and breast. Besides that, I made
cold-water applications to his head and put blisters on his neck. In
about twenty-five minutes he recovered consciousness. I found that he
was taken by hemiplegia. His left arm and left side were paralyzed, including the face and tongue, and each respiration produced a puffing of the
left cheek.

The muscles of the tongue were affected also, (the hypoglossus nerve
being paralyzed,) so that when the patient was requested to show his
tongue and he did so, the point would be deflected toward the left side.
I made him take purgatives. I gave him a cathartic consisting of castor-oil and three or four drops of croton-oil. This operated upon him
three times, not to any great extent, however. He had not eaten much
during the time he had been out. On sledge-journeys you have to try
to save your provisions. He slept some hours during the night. Mr.
Morton kept watch at his bed. On the morning of the 25th he took
some arrowroot for breakfast, but he experienced some difficulty in swallowing it. He complained of the numbness of his tongue. Sometimes
he was entirely incapable of speaking distinctly. Again I gave him a
dose of castor-oil and-croton oil, and he recovered from his paralysis
pretty well. On the 26th he had a restless night, and hardly any appetite in the morning. He asked for arrowroot, but when it was ready he
would not take it. He ate some preserved food. I think he took some
peaches and perhaps some pine-apple, but I am not quite sure as to
that. He complained of chilliness, and indeed he had some very rapid
changes of temperature—changes of temperature like you find in cases
of intermittent fever. I tried the temperature by a thermometer. I
applied it to him. The temperature sometimes rose to 111° and fell to
83°. I applied it in his arm-pit and sometimes in his mouth. He did
not like to have it applied to the arm-pit. His temperature was higher
in the evening. This was on the 26th.

Question. What was the state of his mind at that time?

Answer. The state of his mind was as well as ever before—quite clear
at that time. Before that he had shown no symptoms of delirium whatever, nor was he delirious after that—at least I would not call it delirious. He regained his intellect entirely after he had been in this comatose condition. After having experienced these sudden changes of
temperature, and he having recovered from his attack of apoplexy, I
gave him a hypodermic injection of about a grain and a half of quinine
to see what the effect would be. There was a decided intermission, as
shown by the thermometer, and for that reason I injected a small

dose to see what effect it would have. He felt better in the evening. His temperature was normal. He took a little arrowroot and some soup. On the 27th his appetite improved, but he complained again of numbness in the tongue. He experienced difficulty in speaking. On the 28th he showed the first symptoms of a wandering mind. I saw him in the afternoon, and at 3 o'clock he jumped out of his bed, supposing that Captains Buddington and Tyson were after him with a gun to shoot him. I told him there was nothing of the kind, and sent for Captain Buddington and Captain Tyson. Captain Buddington came, and he seemed to be satisfied, but during the evening he grew worse and worse. He accused everybody. He thought that the cook was after him to shoot him, and gave one spring forward with a knife. He examined, if I am not mistaken, Captain Buddington's mouth, and said that he saw blue gas coming out of it, and thought they wanted to poison him. Mr. Chester wanted to give him a pair of stockings, but he would not take them for fear of being poisoned. He labored under such hallucination during the whole day. He was apparently well, but he did not take anything except canned food, and he opened these cans himself so as to be sure not to be poisoned. He was strong enough to do that. If he did not succeed in opening them he would have one of the natives assist him. He would call upon his Esquimaux, or Hannah, to do it for him; and during that time Hannah and Joe were the only persons that attended him. He did not trust anybody else. Morton was sometimes with him, and one of the men afterward, and he made them taste everything he took; even the food he took out of the cans. That state lasted until Saturday, the 4th of November. He would not allow me to go and see him from the 29th to the 4th. I did not have him under treatment during the whole time. He had some pills, and different medicines in a little box, and he took them. I do not know how many he took. He always wanted some pills. He asked me several times for pills, and to satisfy him I made him some pills of bread, and gave them to Hannah to give to him to take. He thought they did him some good at the time. On the 4th he grew more reasonable, but then there was a great difficulty in his speech. Sometimes he could hardly move his tongue. He complained of the heaviness of it, numbness, and sometimes in asking questions he could not give a decided answer, and hesitated considerably. That was both from his inability to articulate, as also from want of words. His paralysis on the left side was nearly gone, except so far as the organs of speech were affected. In requesting him to show his tongue he would do it, but from the time of the first attack the tip was always deflected toward the left. I bathed his feet with warm water and mustard on the 5th, and I tried to do it again on the 6th. He thought I was going to poison him with the bath, and I thought it was better not to excite him too much, and so I left him alone. At 1 o'clock on the morning of the 7th he jumped out of his bed, asking for Captain Buddington and Hannah. I was at the observatory at the time. Mr. Chester sent Mr. Meyer and Noah Hayes to call me. When I came he asked for some water, and, on examining him, I found that the pupil of his left eye was dilated and the right contracted. After having taken some water he went to bed. When I asked after the state of his health, he said that he felt rather worse than he did the day before; that he experienced more difficulty in speaking. He became comatose, and, at the same time, as soon as that happened, you could hear gurgling or râle in his throat; and, of course, under the circumstances, I could not attempt to bleed him.

By Surgeon-General BARNES:

Question. Did he become gradually comatose, or was there another sudden seizure like the first?

Answer. There was another sudden seizure like the first. His left side seemed to be paralyzed again. Previous to this, on the 2d day, he had regained his power of motion on the left side, and had apparently entirely recovered from his paralysis, except in the tongue. He now seemed to be paralyzed again on his left side. I found that out by trying him with a pin to see if there would be any muscular motion, but there was none. I tried the right side also. There was a sensation on the right side, but apparently none on the left. Finally we noticed reflectory or spasmodic motions of his muscles on the left side, resembling Saint Vitus's dance on one side of his body. Occasionally the same symptoms were noticed on the right side, but very seldom, and to a much less degree. This was on the 8th. At 3.25 in the morning he died. I did not attempt to bleed him after I heard the rattling in the trachea which I have described.

Question. Give us your opinion as to the cause of his first attack.

Answer. My idea of the cause of the first attack is that he had been exposed to very low temperature during the time that he was on the sledge journey. He came back and entered a warm cabin without taking off his heavy fur clothing, and then took a cup of warm coffee, and anybody knows what the consequence of that would be. I did not look at the thermometer when I entered the cabin to see what the temperature was, but I found the room very warm; so oppressive that I opened the door before I went to his bed.

Question. What had been his physical condition before he went on the journey? Do you know anything about that?

Answer. Sometimes he used to complain of a headache, and of a numbness of his hand, or some part of his neck. He did that after we left, and I did not take it as a very good sign. Before he left on this journey I noticed nothing in particular. He appeared to be in his usual health. When I first came to him, after his first attack, I asked him how he had been during the last days of his sledge-journey, and he said that he had not felt quite well; that he felt a weakness in his legs, and sometimes suffered with a headache.

In the cabin in which he was when he was first taken sick there were eight berths. It had about 1,000 cubic feet, I should say. It was about 15 feet long and about 8 feet wide. Seven people slept there, including the captain. They all slept there during his sickness. The ship was housed in with canvas and banked up with snow, with a narrow passage-way at the gangway to come in. The change was very great, coming into such an atmosphere from where he had been on that journey for two weeks. He had been exposed to temperature as low as 20° and 25° below zero. His coming into this cabin, where the temperature was so different, produced a sudden reaction. The temperature of that cabin was from 65° to 70°.

Question. What medicine did you administer to him during the course of his sickness?

Answer. Some castor-oil and croton-oil, and some citrate of magnesia. During such intermittents I gave him an injection of sulphate of quinine. That is all the medicine I gave him. In fact you could not give any medicine in a case like that. I used mustard applications, and applications of cold water, and put a blister on the back of his neck. With regard to his appetite, he had to keep dieting all the time.

G P

He wanted to eat seal-meat, &c., but I could not let him have it, and for that reason he accused me of wanting to starve him to death. At one time he got Hannah to cook him some seal-meat, and I could not prevent him from eating it after she had done so. I think he ate quite a lot of it on that occasion. That was Saturday, the 4th of November; it was the day he grew a little more reasonable.

Question. Do you remember his refusing to take medicine and Captain Buddington saying "Mix up rather more than he wants, and if he sees me take a little of it, he will take it?"

Answer. Yes, sir; he said so. He had not had a passage for two days, and I wanted him to take some medicine, and I could not get him to do it. So I mixed some castor-oil with croton-oil again, and gave it to Captain Buddington, and requested him to give it to Captain Hall to take, but Captain Hall would not do it. I could not get him to take it in any way. I gave it to Hannah, and he would not take it from her. He asked for some cathartic pills. I gave him some of those. When he was given some he buried them under his pillow. After his mind began to wander he grew very suspicious of everybody. He thought everybody was trying to poison and murder him. He never showed any direct suspicion or made accusations against any one before his mind began to wander. He did it the first time on the 28th of October. The first day the mustard applications were made I made them myself. Hannah brought the mustard to the cabin, and the steward brought the warm water, but I mixed them and applied them. Captain Buddington saw his tongue deflected, and Mr. Meyer saw it, and I think the Esquimaux Joe and Hannah also saw it. If I am not mistaken Chester and Morton saw it also. Morton was with him during the greater part of the time.

By Surgeon-General JOSEPH BEALE, U. S. N.:

Question. Did Captain Hall have any stertorous respiration, or did he breathe quietly?

Answer. He breathed quietly; there was no stertor at all.

Question. How did you know in the first instance that the first attack, that lasted 25 minutes, was not a case of syncope? You call it a comatose condition. How did you ascertain it was not a case of syncope? Might he not have fainted?

Answer. O, he was paralyzed.

Question. How did you know he was paralyzed? He was lying in his berth?

Answer. Yes, sir.

Question. How did you ascertain he was paralyzed? Was it a paralysis both of motion and sensation?

Answer. It was only paralysis of motion after the recovery. His paralysis did not leave him until the next day.

Question. Motion and sensation both?

Answer. Yes, sir.

Question. Did you try the sensation in the first attack?

Answer. Yes, sir; I tried it with a needle.

Question. How did you try the paralysis of motion?

Answer. I lifted his hand, and as soon as his hand was lifted it would fall.

Question. You had no doubt, then, that it was a case of that kind?

Answer. O, no, sir; there was not the least doubt about that. As soon as his hand would be lifted it would fall back again. He was not able to support it.

83

Question. You have mentioned that there was an interval of four days during which you did not attend him professionally.

Answer. No, sir.

Question. Did you see him during that time?

Answer. I saw him in the morning before I went to the observatory, and in the evening before I went to bed.

Question. Was there any medicine administered to him?

Answer. Nobody gave him any. He had some in his drawer. I examined it after his death. I found there some cathartic pills and some patent medicines. I found no narcotics, no opium.

After Captain Hall's death Captain Buddington and myself held a consultation, the result of which I suppose you have seen. I can produce the original. It was put down in black and white, communicated to the officers of the ship, and, if I am not mistaken, it was copied into the log. It was signed by both of us.

(A paper was handed Dr. Bessels, which he recognized as the original statement taken down in his own handwriting. It was marked by the Secretary, "No. 3, E. B.")

Dr. BESSELS, (resuming:) I am really at a loss as to what to say of what occurred during the winter. Observations were kept up diligently—meteorological, astronomical, and magnetic. We had two snow-houses connected with the observatory, one of them containg a declinometer, and the other a dip-circle. The tidal observations began on the 6th of November, 1871, and continued until the 7th of June, 1872, comprising nearly eight lunations. These observations were kept up hourly, sometimes half-hourly, and to establish accurately the turn of the tides at intervals of every ten minutes, I compared the observations every evening, and had supplied the service with a good time-piece. These observations proved the important fact that the tide of Thank God Harbor is not produced by the Atlantic, but by the Pacific tidal wave. It was found that the cotidal hour is about $16^h 20^m$. Rensselaer Harbor, being the northernmost station, has its co-tidal hour at $18^h 04^m$, consequently the tide comes from the north, the rise and fall at spring-tides amounting to about five feet; at neap-tides, $2\frac{3}{10}$. Sometimes we had opportunity to determine the velocity of the current. Once we made fast to an iceberg, and by means of a log-line I measured the velocity of both the ebb and flood current, and I found the velocity of the flood-current to amount to more than that of the ebb; and sometimes the flood will continue to run while the water is falling. The iceberg was aground. The ship was made fast to it, and I hove a log-line and maul-line fastened to one end of the log-line. I threw it on a piece of ice, let it run out, and noted the time elapsed. I have about ninety-two measurements of velocity. Most likely the two tidal waves meet somewhere in Smith Sound, near Cape Frazier. Kane and Hayes have both found a ridge of hummocks near Cape Frazier, and in drifting down we experienced that during some time, being abreast of Cape Frazier; we hardly made any headway, but we drifted both north and south.

Rensselaer Harbor is the northernmost point known where the Atlantic tidal wave touches, and consequently both of those waves must meet somewhere. I suppose the tide we have at Thank God Harbor is the Pacific tide. We might call it the polar tide, because Behring Straits being very narrow, it is hardly possible that the tide can originate there. It was to the eastward of Spitzbergen, and between Spitzbergen and Nova Zembla, I noticed that two tides meet. I was there in 1869 with the German expedition. I wrote to the commander of the last Swedish expedition, at Spitzbergen, to send me some notes of his tidal observations up there, they being the northernmost observations, except ours, that have ever been made. We have hardly datas enough to understand the tides until now, because there had not been observations enough in existence, but I think we are getting so now that we may be able to prove that the tide is really the Pacific tidal wave. During the winter we kept up the scientific observations. I have copies of those observations here. I have two books of tidal observation. (Dr. Bessels produces four books, two of tidal observations and two of meteorological observations.) These are original records of observations taken at the observatory, at Thank God Harbor, in latitude 81° 38'. I had, besides, some others, but lost them when the ship went adrift on the ice, as I shall hereafter detail.

After the appearance of the sun in 1872, I handed in a plan of operations to Captain Buddington.

Dr. Bessels being shown by the Secretary a paper marked "No. 5, B," he says: "This is the original paper, in my handwriting, which I handed in to Captain Buddington." Being shown by the Secretary paper marked "4, B," he says: "This is the letter which accompanied it."

Captain Buddington wrote me a letter to the effect that most likely the final expedition toward the north had to be made in boats. I have the original letter in Mr. Meyer's handwriting, signed by Captain Buddington, and his reply.

(Paper produced and marked by the Secretary, "No. 5, E. B.")

It being of the utmost importance now to connect Kane's farthest point with our survey, Mr. Bryan and myself started on the 27th of March for Cape Constitution. We had a sled with eight dogs, and Joe as driver. On the evening of the same day we arrived at the sound called on the chart prepared by Mr. Meyer the Southern Fjiord. We encamped on the island near the northern shore of it, and proceeded to the interior of the fiord on the morning of the next day, penetrating about twenty-eight miles, when our progress was checked by heavy icebergs that had accumulated. We could get no farther. We staid there to take some observations; fixed our position and made surveys in the vicinity. Besides that, we took a sounding in one of the tide-cracks, not getting any bottom at a depth of ninety fathoms. The next day we left and encamped again on the island.

When we undertook to start the next morning our sled broke down, and we had to send Mr. Bryan and Joe back to the vessel to have it fixed. I remained there until they returned, which they did in about thirty-six hours.

We succeeded in crossing the fjiord. We traveled along the western coast of Greenland, where we found at a distance of about thirty miles south from the fiord another deep inlet, which was explored. Said fiord is not marked on the chart made by Mr. Meyer. The track marked is not quite in accordance with the truth. We encamped again on a little island abreast of said inlet, and the next morning Mr. Bryan, Hans, and myself started to look for Cape Constitution, and supposing that we would find it in doubling the south cape of the islet. We found ourselves disappointed. In fact we could not see anything like Cape Constitution. So we had passed the latitude of the said cape, as indicated by Kane's map. I mean the second edition of Kane's map. There is a difference of about twenty or thirty miles in latitude between the first edition and the second. Kane took the mean between his dead reckoning and the actual astronomical observation; consequently his positions are so much farther to the north, because an arctic traveler is very apt to overrate the distance he has traveled. We traveled during fifteen hours, finding the ice extremely rough. We had to abandon our sled, and climb over steep cliffs, there being no ice at some places, the water touching the rocks immediately. After some time we saw some smooth ice ahead, and thinking we were able to make some headway we turned back and carried the sleds and dogs over the cliffs. After having proceeded about ten miles farther we were arrested by open water. We could not reach Cape Constitution, nor could we see it plainly, but we noticed an island at a distance of about 25 miles to the southward. Morton, who had been with Kane when Cape Constitution was discovered, pointed to said cape and told me of such a place, where Kane and himself had been some years ago, and where they had killed two bears. We saw this cape plainly after we came down and identified it clearly as the same place—all of us, Hans, Morton, and myself.

Dr. Bessels produces sketches of Cape Constitution and the vicinity with the croquis of the rough survey. The papers designated, respectively, 1, 2, 3, 4, 5, were placed in an envelope and marked by the Secretary, "No. 6, E. B."

Those sketches were made on board the ship on our way home. I had some other sketches of Franklin Island and Cape Constitution, but they were lost—sketches that I took when I was out with Mr. Bryan, and Joe, and Hans.

We continued our travels until stopped by open water. We could not fix our lost position for astronomical observation, because it was cloudy and we had to make it up by dead reckoning. If we put Cape Constitution in 80° 25' I think it will do justice to Kane. I think that is as high as it can be made. Perhaps you can make it thirty minutes higher, but not more. The land continues on to the northward above Cape Constitution to a considerable distance, instead of there being an open sea to the north of it.

We found an open sea there as Kane did; but open seas do not amount to much, because they are merely local, that is, the water is kept open by the influence of tides and winds. Sometimes the velocity of the wind amounted to sixty miles; there was a strong tide, and I do not think any ice could withstand that. Besides, the cliffs are nearly perpendicular, and no shore-ice can form there. Consequently the water has to remain open. There is no open Polar Sea there. Cape Constitution is on a narrow channel, and the land runs to the north of it for a considerable distance, instead of breaking off there, as is shown on the chart of Kane and Hayes. Cape Constitution is not higher than 80° 25', and above that the land continues to the northward at least forty miles, before we come to the southern cape of the inlet, now called Southern Fjiord, making the eastern shore of a channel or sound, instead of an open Polar Sea, as was supposed by Kane from Morton's account.

On the 8th of April we returned on board the ship from the sledge journey. Several smaller expeditions went out for hunting and surveying purposes, until finally, on the 7th of June, we could make an effort to start with our boats. It was utterly impossible to proceed to the northward by means of sledges, the ice being too rough, there being too much snow, and the condition of the land not permitting of any travel—the configuration of the land. I find in Captain Tyson's statement that he thought it possible to travel overland, but the whole amount of atmospheric precipitation measured during our stay at Thank God Harbor was not more than $2\frac{1}{10}$ inches. I deduce from that that there was not snow enough. The greater part of the land was nearly bare of snow through the winter, and you would have been compelled to draw your sledge over it. There was deep snow accumulated in different places, but the greater part of the land was entirely bare of snow. At St. John's—341 hours, all told, during the time of our sojourn—there were .214 of snow there. The melting of all the snow that fell amounted to a little over 0.2 inches of water.

On the 7th of June Mr. Chester left with his boat to proceed to the north, but unexpectedly returned on the 9th. I did not go with him. Mr. Meyer and four of the men went with him. He stated that he had lost his boat in the drifting ice. The next day Captain Tyson, four of the men, and myself went with another boat, and we were fortunate enough to get by Cape Lupton, where the tide runs rather swift, reaching the middle of Newman's Bay on the evening of the same day. On the 14th, Chester joined us in the canvas boat with his whole crew, and I staid there until the 1st of July, without having been able to proceed any farther. The ice kept pouring down during the whole time through Robeson Channel. There was not water enough to float the boat, and the ice was not solid enough to travel over it. It consisted of small pieces during the first time and in hummocks, and only toward the end we had heavy fields coming down. In fact Robeson Channel was not frozen during the whole winter. It was always on the move, except during a few days in March. In the consultation between Captain Buddington and myself, you will see that we intended to start a party to Grinnell Land, but were prevented from doing so by the breaking up of the channel the day before I transmitted to him .this paper. On the 1st of July, Captain Buddington sent a message by Hans, and said our presence on board the ship was required. I returned immediately with Hans, finding in making the south shore of Newman's Bay the first piece of drift-wood, and that is the only specimen preserved. It is just as I found it. I merely made a section of it, and polished it a little. Most likely it grew somewhere where there are extreme temperatures, very high temperatures during the summer and low temperatures during the winter. It is about twenty-seven years old. I cut it, and here is a piece belonging to it. After I had found it, I looked for about eight hours, and Hans did the same, but could not find another piece until we arrived at a plateau from about sixteen hundred to eighteen hundred feet high; and there we found this drift-wood, all consisting of small pieces, most likely Siberian pine; but it is difficult to identify it, although not impossible. It can be identified. The reason why I think the first piece I found is about twenty-seven years old is from the annual-growth points. It has twenty-seven, more or less, distinct rings, which can be seen and counted under the magnifying-glass. I think all the pieces of wood are Siberian pine. I can hardly think any walnut has been found.

Question. Did you see the other pieces?

Answer. No pieces have been brought down except the few pieces I found.

Question. Those are the pieces Mr. Meyer refers to when he speaks of the wood as being walnut?

Answer. This looks a little like walnut, but it is no walnut. He states it smells a little like walnut, but I do not think walnut has any specific smell. These are the only pieces I saw.

In coming on board ship I learned from Captain Buddington that he had tried to push on to the north, but had been forced back by drifting ice; that he had attempted to take us off at Newman's Bay. He said that he had fired guns and made different signals to attract our attention, but we never heard them. Mr. Chester sent two of his men to get some provisions. He intended to spend some more time up there, but finally he and Tyson were compelled to return because there was no prospect whatever of getting any farther north. If any other pieces of driftwood were found in Polaris Bay I suppose they were found by seamen who saw them without knowing or appreciating their value, and did not bring them on board the vessel.

These pieces which I produce are the only pieces of drift-wood that I saw on board the ship. I have referred to the land to the north of the northernmost cape of Grinnell Land without any name. On the 7th of August two of our seamen, Robert Kruger and Henry Hobby, went back to Newman's Bay to get some of their clothing, and in going there they saw the land as plain as it possibly could be. One of them, Henry Hobby, remarked that the northernmost cape of Grinnell Land seemed to be so near to him that he used the expression that he could "spit on it," and he described the land to the north of the cape as perpendicular steep cliffs, covered at some places with snow; and the account of the land corresponds exactly with the bearings of the cloud that I had taken some weeks previous to that to the northeast of where he was. That is above the northernmost cape of Grinnell Land. This land lies above the northernmost point of that unnamed region which lies above Lady Franklin Bay, as laid down on the chart of Mr. Myer, and seems to be disconnected from it and lying off to the northeast and trending from northwest to the southeast. Now, Mr. Meyer, when he was at Repulse Harbor, stated that he could see a shining spot, and he took it to be open water. Now I do not deny that. I merely said that there must be some land behind such open water, and that in consequence of the open water he could not see the land. This is not uncommon. During our second winter-quarters at Polaris House we hardly ever had an opportunity of seeing the land which was opposite to us, though being only thirty miles distant. Sometimes we could see it. As an illustration of what I mean I will state that at our second winter-quarters at what we called Polaris House, at Lifeboat Cove, generally we could not see the land across toward Cape Isabella, on account of open water which lay between; but as soon as we had heavy southwest winds so that the water was blocked up with ice, and the frost-smoke from the water prevented from escaping, we could see the land plainly, and so I supposed this to be the reason why Meyer did not see this land which lies at the north. Two men saw the land quite plainly without glasses. Mr. Meyer had a glass. I can only account for it from the fact that there was, perhaps, an unusual amount of refraction at that time.

There is a great deal of refraction up there, and it is a refraction that is very unequal. We saw a great many mirages during the whole time. Sometimes the land seemed to be lifted up a great height, much higher than it was in reality, and we could see it actually at a great distance. We could see objects really situated below the horizon. This refraction was very frequent in these latitudes, and very unequal, as I have said. I should say this land which we saw was not farther south than $84° 40'$, and is, of course, far the most northern land ever seen by human eye. I proposed to call it "Grant Land," as being the most northern land that we discovered, or that has ever been discovered, and shall so mark it on the chart to be prepared by me, knowing that it was Captain Hall's intention to name the highest land discovered after the President.

The water-cloud hanging over the open water in front of this land was seen by Mr. Meyer and also by Captain Hall, as he states in his dispatch, and was seen by us, when we were on our boat-journey for several weeks, daily. I took the bearings of it, and the land which was seen at intervals, by Buddington and myself in 1871, and afterward by the two seamen, as I have mentioned, lies behind this cloud and could only be seen at intervals, being obscured by the frost-smoke from the open water in front of it.

Without concluding the examination of Dr. Bessels, the commission adjourned to meet to-morrow morning, at 11 o'clock.

WASHINGTON, D. C., *October* 17, 1873.

Examination of Dr. EMIL BESSELS resumed:

During the whole time of the boat expedition to Newman's Bay, the ice poured down from north to south with the exception of twice. On those occasions it went to the northward, except when it was at the time of the spring-tide, so that most likely it was under the influence of the tide. But the motion never lasted long. The first time it lasted half an hour, and the second time about fifteen minutes. It moved against the wind. The wind was from the northeast, with a velocity of about ten miles, and the ice moved to the northward against the wind. I came back on board the ship and found her leaking worse than before. The next morning another attempt was made to go to the northward in the ship, but we did not get very far. We did not even reach the latitude of Newman's Bay, and were compelled to go back; but, previous to that, we landed two men belonging to Mr. Chester's boat carrying some provisions. We came back to our old anchors, and during the whole time, until we started, we were troubled by some moving ice. We had to move nearly every day, and very often it happened that the ship grounded.

There was no possibility of stopping the leak. Captain Tyson states that there was rise and fall enough to do it, but in reality there was not. As I mentioned yesterday, the rise and fall did not amount to much, and, if I remember aright, gave it in figures. We could do very little during that time. We were not able, even, to get another set of pendulum observations, which were very desirable, because we had to be on the move during the whole time we lay there. Sometimes we moved three and four times a day, simply for the purpose of keeping clear of the ice. The different parties returned from the boat expedition; first Tyson and then Mr. Chester; and on the 12th of August we had to bear up for home, the ship being in a leaky condition. We had not coal enough to stay another winter, and to steam down the next year. It is my opinion that if the ship had not been leaking, and if we had had a whole ship, we could have staid there and continued our researches. After having passed Cape Constitution coming down, we got beset. In coming down we noticed Cape Constitution. We saw it distinctly. Those sketches that I handed to the Secretary yesterday were taken abreast of Cape Constitution during our voyage down. Both Morton and Hans recognized the place. Hans recognized it when he was with us during the spring, on the sledge-journey. He drew off an outline of the coast in the snow. I did not trust him at first, because I had left a chart in the snow-house where we encamped, and I thought that Hans might have seen such chart, but his answers to our questions were of such a positive character that we were disposed to believe him. He pointed out the spot where Mr. Morton and himself had killed two bears. I gave the latitude of Cape Constitution yesterday. We could not determine it by actual observation, but everything points to the fact that it is not any higher than 80° 25′. We had sight-lines twenty miles distant from the cape. We were distant about twenty miles to the north of it, and we took a right tangent to the extreme cape, supposed to be

Cape Constitution, from the point from where we stood—north 3° 15½' east, and we took a right tangent to Franklin Island at the same time, north 3° 28' east, and we took the left tangent to the same island, north 2° 18' east. It was not possible to determine it by actual observation, because going down there we had no sun. We had only to fix our last position, and in making our way down with the ship we could not steer a steady course, so we were unable to fix it by true bearing or any other. We estimated the distance, and I marked the latitude of the cape on such sketches as I handed to the Secretary yesterday.

South of Cape Constitution, as we came down, we got beset, and drifted down along the west coast. That is the east coast of Grinnell Land and the west coast of the channel. During the greater part of the time we saw water along the coast, but we were never able to reach it until the catastrophe happened upon the 16th of October—until we broke out. We saw water along the west coast of the channel, but we could never get to it. We were beset in the middle of the channel about fifteen miles off the coast, and then we drifted in the same meridian south until we came to Force Bay. Then we followed the trend of the coast. The last point we sighted was Gale Point on the west coast, when a heavy southwest wind sprung up and there came on a heavy snow-drift. I will state that we saw Cape Alexander and drifted to the southward of that. At 6 o'clock on the 15th of October, in the evening, the ice separated at the stem and kept on separating until all the ice on the starboard side of the ship had gone. After some time the ice set in again. There was considerable pressure on the ship; sometimes she was strained and keeled over on her port side. Captain Buddington ordered the provisions and stores, whatever we had, to be thrown overboard. Nineteen persons went on the ice, partly to help to carry the provisions back to the house that had been erected some time ago in case of emergency. We staid on board of the ship handing and throwing the provisions over. It was about half past 11, if I remember aright, when the two hawsers parted, and we drifted at the rate of about ten knots before the wind and lost a floe with our men out of sight. Previous to that we could see that the floe had broken to pieces, and that one piece, with provisions, coal, and some records on, had gone one way, and the piece with the men in another direction. The third piece had on it two boats and some of the children asleep. Some of the men tried to launch the boat and made for the lost floe. That is the last we saw of them. It was all done in a very short time. It did not take over ten seconds. All the records of Captain Hall and of the astronomical and magnetic records were thrown over the side of the ship. There were several diaries. Part of my papers and the whole of Mr. Meyer's papers were put overboard. I could hardly tell who it was that put them overboard, it was done in such a hurry. I know they were put overboard, because I helped myself to take some of the boxes out of the cabin, and I saw a large Japan tin box belonging to Captain Hall, and containing his papers, which was put overboard. I do not remember exactly who did it, but it was done by either Mr. Bryan or Mauch. It was put on the ice, at all events.

Question. How does it happen that these records of yours were not put on the ice?

Answer. I had one drawer and a box of papers and specimens, and I carried those on the ice, and when I put them down a squall took some of them away, and I covered them and went back on board the ship and put the rest and those I have here in my blanket. I wanted to keep

them with me and then jump overboard with them at the last minute.
I did not, however, get any chance to do it, so I have really saved them.

By the SECRETARY:

Question. But you know that the tin box in which Captain Hall's
papers were was put over on the ice?

Answer. I am quite confident of that; also some magnetic, astronomical, and part of the meteorological observations; also the books regarding natural history, geology, &c.

Question. All of them?

Answer. Not all; only some of them. We saved the pendulum
observations, but the observations of time are lost. We saved a part
of the meteorological observations, the tidal observations, and some other
notes. We saved some specimens, of which I have two boxes here. The
specimens consist of one package of phanerogamic plants, one package
of paleontological specimens, then a collection of insects, one bird, and
some invertebræ and marine animals. That is about all. Besides that,
we have Esquimaux relics found near Polaris Bay, and drift-wood. We
saved tidal observations, as I stated, and they are most complete. Besides that, I left a duplicate of the tidal observations at Lifeboat Cove,
with the log-books.

Question. You threw no specimens on the ice?

Answer. O, yes, sir; we threw our collections over. There was a box
of nearly all the stones thrown over, all with the exception of two boxes.
There was a barrel of bones, and another barrel of skins—all the rest
of the dry skins and some bones. Those all went on the ice and were
lost, except some skeletons. We had not put up any barrels of those,
and so we have five or six musk-ox skeletons and a number of squirrels.
We had a complete series of musk-ox skeletons. We had some sixteen
skeletons, from the fœtus to the full-grown animal. Unfortunately we
could not save any of those. We had a part of them on board the ship
and took them on shore. We had no room in the house to stow them
away, and the Esquimaux took the horns and used them. It would
have been entirely impossible to have carried them with us in the
boats.

On our return we were carried by the influence of the wind. The
water was gaining fast on the ship. We tried to start our deck-pumps,
but found them frozen in. Finally we succeeded in working them, but
still the water would not diminish. It nearly threatened to extinguish
the fire under the boiler. Finally, by smashing the doors and heaving
blubber into the fire, we succeeded in raising steam enough to diminish
the water, and some of us turned in, and after daylight we found ourselves somewhere near the coast. We took one of the sails down and
cut it up to make bags and put our coal in, and at about 8 o'clock in
the morning we began to work the ship in-shore, there being a lead just
leading to the shores of Lifeboat Cove. In the morning we found ourselves abreast of Lifeboat Cove—just abreast of that lead. We
worked the ship under canvas, and steamed as well as we could without the assistance of boats, and managed to get in, and used two tides
to get up as high as possible. Then we landed our stores, and Mr.
Chester, with the carpenter and some men, began to build a house for
winter-quarters.

We kept a constant lookout on the mast-head for the men who were
separated from us, but we never detected anything that looked like
them; once Mr. Chester thought he saw a piece of ice with some barrels and bags on it, but opinions differed in regard to that point; and

even during the night we tried to make signals for the men. We had a lantern at our mast-head, and we put it up three times, but the wind blew so violently that it was extinguished with every attempt that was made to keep it lighted.

Question. How do you account for the fact that the men on the ice saw the ship so distinctly, and that they were never seen from the ship?

Answer. I really cannot account for that fact. It may be that they mistook an iceberg for our ship. That is often the case. Men often in that region mistake an iceberg for a sail. The fact is that we never were as near Northumberland Island as they state they saw us. At first, before we made the shore, we thought we perceived some men at Lifeboat Cove, on shore. But finally those objects we took to be men turned out to be bowlders. We kept a constant lookout for them. It is possible that they might have been in the shadow of some iceberg or the shadow of some hummocks.

Question. Could your vessel have been lifted up by the mirage so that they could see it when really it was out of sight?

Answer. It is not impossible that it might have been a case of mirage in the northern horizon, not in the southern. They might have seen us when we were not able to see them. It is also possible, as I have stated, that they might have been in the lee of some iceberg or hummock, or have been so near the shore so that we could not see them.

Question. How far north do you think you were blown on that occasion before you returned, after the hawsers parted?

Answer. About twenty-five miles, I think, we drifted that night. That is only an opinion. I would not like to state that positively.

Question. How far southward did you make again before you went into Lifeboat Cove?

Answer. We hardly made any. The only lee we could see was abreast of the ship, and we stood in for that lee.

Question. At the time these men saw you, then, you must have been nearly twenty-five miles off from them?

Answer. O, yes, sir; but perhaps not. It is possible that the ice-floe, being smaller than the ship, drifted faster. We moved about five miles that day.

Question. Was there any mist on the water at any time that might have prevented you from seeing whole objects in the southern horizon?

Answer. It was clear. The gale was over. The gale had abated about half an hour after midnight, and at that time we could see the moon. It was nearly full then; it gave considerable light, but we could not see anything. We could just see the dark outlines of the land; that is all. We could make out where we were. When we reached shore at Lifeboat Cove we landed as much of our provisions as possible, and Mr. Chester and some of the men set to work to build winter-quarters. The next morning some Esquimaux, with their dogs and sleds, came, stating that they smelt the smoke of the ship, and offered their services. They said they had not seen it. They made statements that they smelt it.

Question. Is their sense of smell very acute?

Answer. I should not think it was. They offered their services, as I stated, and we were very glad to accept them. These Esquimaux came from Eta, about twenty miles from the south from where the ship was. They said they smelt the smoke there. We had a light breeze from the northeast, so they had exactly the wind that would enable them to smell the smoke if it were possible.

Question. Might not they have seen the vessel from Eta?

Answer. No, sir; they said they smelt the smoke, and that their dogs smelt it. As I have stated, the wind was blowing in the right direction for them to do so. We had a light breeze from the northeast. We did not take any observations; but I am certain the wind was from the northeast at the time. After the house was done and we were made as comfortable as possible, we set up an observatory on shore and took as many observations as our instruments and means would permit us. All the magnetic instruments had been lost. Some of the instruments had to be manufactured to take the place of those—some meteorological and astronomical instruments. The instruments that I saved all were lost on the ice. We lost, among other things, our declinometer, our dip-circle, and one box of chronometers—our standard chronometer. We lost one pocket-chronometer—Parkinson & Frodsham—and some barometers, and so forth; also a box belonging to the photographic apparatus exclusive of the camera.

During the winter we took such observations as our instrumental means would permit us. Astronomical and meteorological observations were made—the latter hourly. We saved mercurial barometers and aneroids, a number of thermometers, anemometers, psychrometers, and some other instruments. One of Regnault's actinometers was made to determine the temperature of space. We lived pretty comfortably in our house, only the ventilation was not very good except during heavy gales. We had a great number of them, and could hardly keep the house warm. Our bunks were lined with ice; the ice accumulated everywhere. We had nothing but a light canvas roof over the house, but, fortunately, it was improved by lining the inside with some old timbers of the ship. Our coal did not last us very long, so we had to take to the ship and burn her up as economically as possible with our fires. On various occasions we did not have any fire during the night, trying to economize fuel. Once we tried to see how it would do to cook in the house instead of the galley, to see if we could not in that way economize fuel, but we found that we consumed rather more fuel than if we cooked in the galley. The thermometer would indicate, about eight feet from the stove, some seventy degrees, while at the other end of the house a cup of water put on the floor would freeze; so we had to abandon that and take to our old galley again.

The lowest temperature during the winter was some time in March. It was about 44 degrees minus. The minimum temperature was from the 3d to the 4th of March, and that was 37°. We read our minimum thermometers at 8 o'clock in the morning.

Question. What effect did the lowest temperature have on the mercurial thermometer? Did the mercury harden?

Answer. The mercury seems to congeal at about 39.9. It is of great importance never to take thermometers with narrow bore. The thermometers with the wide bore that we had, and which were supplied by the Signal Corps, indicated correct temperatures down to 40, but those supplied by Cassella, in London, with narrow bore, would stop sometimes at 35.

By the SECRETARY:

Question. You made lower temperatures there than you made up in Polaris Bay?

Answer. Our mean temperature at Polaris Bay will be a little lower,

I think. Our minimum at Polaris Bay was 48, and it occurred in January, I believe, 1872.

Question. How did you obtain this—with what instruments?

Answer. We used the mercurial thermometer down to 40, or rather, as soon as the temperature came down as low as 35, we registered both the spirits thermometer and the mercurial thermometer. We registered both until the column of mercury was contracted down until we could not read it any longer. But all our instruments had been compared, at intervals of 10 degrees, from the highest temperatures we experienced down to the lowest, and those corrections had been applied, but still here in the books both instrumental readings have been given. I will now hand you some of the reductions of the different observations.

Question. Your mercury did not harden at 58 so that you could take it up like a shot, did it?

Answer. Our mercury grew hard, quite hard. At Polaris Bay we made some balls, and fired them through an inch plank. If the mercury is not quite pure it will not freeze at 399. Sometimes it occurred that it would remain fluid when the reliable thermometer showed 42, but then the mercury was never pure. Pure mercury seems to congeal at 39.9. That seems to be the point to adopt as a freezing-point of it; at least we did so for our corrections.

By the SECRETARY:

Question. What else did you do there?

Answer. We made all the observations we could make, and tried to get some dogs from the natives, and on the 13th of April we made another attempt to push on north to reach the provision depot at Polaris Bay. It was impossible to do it any sooner, because we had no skins, and had to send one of our Esquimaux—Esquimaux Jim—to the southern settlement to get some deer-skins to make some stockings, and some blankets to sleep on. It was on the 13th of April before we could start. We started with two Esquimaux, this Esquimaux Jim and another Esquimaux, Awatak. We started with those two Esquimaux who were willing to go. I pretended that I wanted to get some musk-ox, and was going to hunt for that purpose. They consented to go. I supplied one of them with a rifle. They were very anxious to go with me to hunt musk-oxen. A short time after we left, a light breeze sprung up, and they began to want to go back to the house, where they had had a comfortable time previous to that. After we had been out four or five hours, they didn't want to spend another night there. I did not agree with them on that point, and pushed on until we came to the hut at Anoatock. We put up in a snow-house that was about thirty miles from the ship. The next morning we started. The ice on the east side being very rough, the natives wanted to make me believe it was too rough to pass over. But I could see a smooth ice-foot along the shore. They persuaded me, however, to direct my course to the west coast. I did it, because it could not make any difference to me whether I made northing on the east or the west coast. Finally, however, I found that one of these natives did not intend to travel with me north, but wanted to go west and south in search of a bride. That did not agree exactly with my intentions, and he grew impudent. He was dissatisfied. He spoiled this man Jim, who seemed to be quite a reliable man. He requested me sometimes to give him something in my possession, and I consented to everything that I possibly could, hoping in that way to be able to go ahead. The highest latitude we reached by observation was 79° 16'.5. That was on the 16th of April, 1873. We ran about thirteen miles and met the east coast of Grinnell Land. There we met a ridge of hummocks. The natives refused to go farther. Jimmy declared that if I insisted on going any farther he would return on foot and go home to his wife. He told me that the dogs belonged to me, that we had given him his sled, but he would let me have it. He said I should drive the dogs and go alone wherever I had a mind to. I succeeded in satisfying them by giving one of them a hatchet and the other a saw, and stated that if they deserted me again I would take the hatchet and the saw away from them. They consented to cross the hummocks, the ridge of hummocks being about half a mile wide. We found three or four miles of level ice, and then we came to another ridge of hummocks. I could not get them any farther. They both wanted to go home, and I thought the best thing I could do was to go home also. In going home Esquimaux Awatak at once turned around on his sled and spoke to me. I could not understand what he meant. Finally he grabbed me by the shoulder and shook me. I did not like that kind of treatment exactly, and I took up my rifle and pointed it at him. The rifle was not loaded, but still he was very docile after that. We traveled for thirty-three hours until we reached the house. The greatest part of the time I was among the ice, making very good headway. I slept a few hours and then took another dog-team and went down to the settlement of Sorfalik to get another Esquimaux. I staid there during the night, and started the next morning with another Esquimaux and eight fine dogs and dog-food. I went up to the house at Lifeboat Cove and took my clothing and some little provisions on the sled, and Jimmy, my former companion, consented to accompany me again. We started. I had left a part of my dog-food and the heavier part of the outfit near Cape Inglefield, and so we shaped our course for that point. This time it was my intention to follow the coast-line of Greenland, though I was compelled to make a great deal of easting. I expected to find smooth ice when we reached Cape Inglefield. The other Esquimaux I got at Sorfalik gave way. He broke his sled on purpose to compel me to go back. I tried to mend it, but unhappily he had purposely left the saw behind that I had given him; so we had to return to the house. I left everything up there that I possibly could, dog-food and all, and we started with both sleds, one being badly broken. We repaired it, however. That was done at 5 o'clock in the evening. Then I wanted to start again, but I could not get those Esquimaux to do it. They would not go beyond Cape Inglefield the second time. They complained that the ice was rough, and that they had so much to do in lifting the sleds over the hummocks, &c., that they refused to go, and some of our party thought that they merely did it because I had pointed the rifle at Awatak. I do not know how that was. The fact is, however, they refused to go. I intended to try it again, but this time with one sled. I engaged one of our men to accompany me. Unfortunately, however, the ice broke adrift, and it was quite impossible to start, so we had to abandon the enterprise. I meant to go up the east coast of Greenland and reach the provisions at the depot at Polaris Bay. We had established a depot of provisions there. We left those things there before we started, because we did not know when we would lose the ship. We left pemmican, canned meat, bread and butter, shot, and such things as that at Polaris Bay. I promised those natives that we would get to a white man's house very soon, and then I would give them plenty of knives if they would go. I promised them a boat that we had left at Newman's Bay, but it was of no use. At one time they would consent, and then they would decline to go.

Polaris Bay is in latitude 81° 38', but if I take those twelve miles into account I went to 79° 28'—about two degrees lower. If they had consented to go on, I think I might have reached Polaris Bay. If I had had Joe with me I could easily have done it, but you cannot govern those Esquimaux; they will have their own way. When Awatak told me that he wanted to wait near a seal-hole and watch the seal, I had to comply with his wishes, because I depended on him entirely. It was different with our Esquimaux. They were civilized and knew a little more about it. The natives did not seem to have any knowledge of the coast north. The natives of the American coast—of the west coast of Grinnell Land, and south of that—have a great knowledge of their country, and they can draw pretty accurate charts. Joe was quite a good draughtsman. We tried it several times with them. I gave them a piece of foolscap paper and they put down three huts at Etah, and put down our house and gave the configuration of the coast-line, and everything of that kind, but they could not give it any farther. The natives of the west coast stated that United States Sound, as laid down by Hayes, is in reality a sound connecting with Jones's Sound, and making Ellesmear Land an island. They call it Kickertack-Soak, which, being translated, means a large island. They at first said so to Captain Buddington, that there was such a sound and such an island. Jim's native name is Ttuckischu, and the name of his wife is Tvallu. They are natives of the west coast. They came up there from Cape Seal. They informed us that Grinnell Land is inhabited south of Cape Isabella, and that there are musk-oxen there, and a good deal of drift-wood, the drift-wood coming from the northward. Before I started I made a survey of the harbor, of the house, and vicinity. I have that survey here.

(The paper produced and marked by the Secretary "No. 7, E B.")

As early as possible Mr. Chester and the carpenter set to work to build the works, and were busy until the end of April and the greater part of May, the weather being rather unfavorable and giving them but little chance to work. They had a very heavy snow-fall during May, and the snow that fell during two days at Lifeboat Cove amounted to more than all the snow during the rest of the time from the second of November up to this time, the whole amount of snow there being 2.31 inches of water more than we got the whole time we were at Polaris Bay.

Mr. Bryan went to Rensselaer Harbor and to Port Foulke to connect the meridians of those places with the meridian of Polaris House, to make it more reliable.

I did not take all the records with me, because we did not know what might happen. Mr. Bryan has a part of the records with him; his diary, astronomical observations taken at Lifeboat Cove, and a view of about fifteen feet long, with all the details, of Grinnell Land. I had it among my papers. I made it going up, but I could not get to it when I left the Ravenscraig. Mr. Bryan, however, took it along and has it among his papers. It gives the details from Cape Frasier. I do not recollect as to the point, but a little farther than Cape McClintock.

Not having succeeded in getting north, I tried to travel inland to the glaciers, but unfortunately the Esquimaux are so superstitious that they fear the glaciers, and I could not get anybody to accompany me. They are afraid of the crevices. The glaciers are intersected by rather deep crevices if you reach a certain altitude, and some Esquimaux in the vicinity of Cape York—I think a man and wife with three children, and a sled with dogs—went down at one time and were unable to recover themselves. That caused them to be extremely superstitious. They think that the glaciers are inhabited by evil spirits, and they declined positively to go. I tried to do it with Jimmy, and the last minute he backed out and told me that his wife could go along with me and drive the dogs. I said, however, that I did not like to take his wife along as dog-driver, so I took Jim. He promised me to go, but unfortunately he only accompanied me to the foot of the glacier in Eta Bay, called by Mr. Kane "My Brother John's Glacier." I staid there for four days and made some measurements of the rate of progress of the said glaciers, accompanied by psychrometrical observations, and tried to ascertain the limit of névé; that is, snow above the ice of the glaciers that has not been converted into ice. The glacier begins as snow, and is converted into ice by packing. We found that the limit of névé begins at elevation of 4,181 feet. I was unable to proceed any farther, because I was alone and the glacier was intersected by deep crevices, so I thought it would be best to return. The temperature proved to decrease 1°.13 F. for every 1,000 feet of elevation. The line of névé is not identical with the snow-line. We did not find the existence of the snow-line anywhere in Greenland.

After some days I returned, and went back to the house. Mr. Chester was still busily engaged in building the boats and arranging provisions for the two boats that had been put up in canvas bags. Finally, we started on the 1st of June to make our way from Melville Bay to the Danish settlement, the observations up at Polaris House being broken up on the last day of May. We started on the 3d of June again. We met with more or less difficulties in coming down, until finally the Ravenscraig picked us up south of Cape York, on the 23d of June. This chart (referring to an ordinary admiralty chart, with certain lines colored by himself) will give you the state of the fast ice as we found it in coming down, the green lines indicating our track until 75°, where the Ravenscraig picked us up. All hands on board the ship were extremely kind. They rendered as much assistance as possible, and took care of our baggage, and brought it on board the ship. Two days after that some of the men went out to take one of our boats on board. They arrived with it after having stove in one of the planks. We were ice-bound without being able to move until the night of the 4th of July, when we bore up and crossed the strait, and went to the westward, at Lancaster Sound. On the 7th of the same month we met the steamer Arctic. Our accommodations being rather poor on the Ravenscraig, Captain Allen was compelled to divide our party, and Mr. Chester, some of the men, and myself went on board the Arctic, Captain Buddington and the rest remaining on board the Ravenscraig. The Ravenscraig party was separated once more, Mr. Bryan and two of the men going on board the Intrepid, and we bore for home. We were unable to reach the Intrepid. We made signals for her, but either she could not understand them or could not get out of the ice. We had not coal enough to go over and take the men, and had to go home without them. We took the men off the Ravenscraig and proceeded to Dundee, leaving Mr. Bryan, Mr. Booth, and Mr. Mauch on board the Intrepid. We arrived at Peterhead, in the northern part of Scotland, north of Dundee, on the 18th of September, and our observations reach as far as Peterhead. The last observation was taken at midnight on the 17th.

Our observations have never been interrupted from the time we left Disco until we arrived at Dundee. We have not complete records of them. The greater part of meteorological observations are entirely lost. We have some valuable observations to prove such a warm current following the west coast of Greenland is not the Gulf Stream, but likely the current produced by melting water trickling over the heated rocks. Sometimes we found even up there at Polaris Bay that the temperature of the water amounted to 54 degrees and more. We tried to use the dredge three times at Thank God Harbor, but the bottom being muddy we did not find any animal life as far as we could go out. And we could not use it frequently on account of the heavy drifting ice.

The whole original survey of Polaris Bay and of the whole coast of Grinnell Land is lost. Mr. Meyer made it.

By Commodore REYNOLDS:

Question. You have now no notes from which that survey can be completed?

Answer. We have no notes except the notes in the log, and some of those croquis with some positions scattered in the journals; but then we can make a pretty reliable map from such data as we have.

By the SECRETARY:

Question. Did you keep a journal?

Answer. Yes, sir; I kept a diary; I wrote it up every day. Unfortunately I used it up the last day at Dundee, in making up an elaborate report. I have such report in a little trunk with those soundings, and some of my private property, which went to London. I saw how it was put in the railroad-car at Dundee, and when we arrived at Liverpool the trunk was gone. I had five pieces, two boxes of specimens, a large trunk, and another valise. Everything was there except that little trunk. I wrote and telegraphed to the consul at Dundee to send it on the next steamer, giving him the address of the Secretary of the United States Navy, and asking him to have it forwarded to him. The railroad officials said it was not lost, and I would get it eventually. It was marked. It had my address on it at Liverpool. It contains three blank-books and a complete journal, written in English.

Question. Did you ever have any difficulty with Captain Hall, except those you have mentioned?

Answer. None whatever.

Question. What was the state of the discipline of ship during his lifetime?

Answer. I could not complain about the discipline; the discipline was good..

After Captain Hall died, Captain Buddington went into command. The discipline after he assumed command was not as strict as it ought to have been. I do not think it was as good as it was before Captain Hall died. I never heard any one, after Captain Hall's death, say that he was relieved by his death. I never heard either any expressions that had that meaning. I think once I heard some expressions before his death which were not very complimentary to him, but that was all. I do not remember who uttered those. Captain Buddington was in the habit of drinking at times. He did not refuse to drink when he could get it. I do not know that he was in the habit of getting drunk, but he was drunk twice, perhaps oftener. Twice I saw him drunk. The first time was during the winter, and, unfortunately, the second time was when we were on our way home with the ship. That was the night when we got off the west shore into the middle of the sound and got beset.

7 P

Question. Did you have any difficulty with him about liquor?
Answer. Yes, sir; a slight difficulty. I knew that he had been getting some of the alcohol. I thought it would be to the interest of the
expedition to take it away from him. Nobody else would do it, and I
was compelled to do it myself. I therefore watched him; I looked
where he went, and he took the bottle—a bottle of alcohol; it was a
half-pound bottle; it was strong alcohol. He got it out of the fore-peak,
out of the scientific stores. The alcohol was kept for preserving
specimens.

Question. As you were coming down you observed constantly the clear
water in toward the west shore?
Answer. Yes, sir; there was clear water along the land. If we could
have kept in there we would have been able to make our way down; at
least I think so.

Question. State why you waited on the east shore rather than on the
west.
Answer. Because the coast north of Lady Franklin's Bay was blocked
by heavy ice and we could not get inshore. Once Captain Hall and Mr.
Chester started across the floe. When we had made fast to it and
made for the land, we found a lead, but unfortunately, however, they
neglected to make the signals or neglected to provide for that. They
came back and informed us, and when the lead had to be tried it was
closed up again.

Question. If you could have made a harbor on the west side it would
have been a great deal better?
Answer. Yes, sir; we could have made a sledge-journey along the
shore. There was hardly any smooth ice in the vicinity of Polaris Bay.
Mr. Bryan, two Esquimaux, and myself went over the whole of the
smooth ice, and it did not extend any farther than from Cape Lupton
down to the northernmost cape of the southern fjord. It was all the
area of smooth ice. All the rest was so hummocky that it would have
been difficult to accomplish two or three miles a day. When Mr. Meyer
went out during the spring, to take some angles for the survey, he was
compelled to leave his sled behind and travel on foot, only to get out far
enough to get sight-lines to some parts of the coast.

Question. Do you think it possible that some other season would have
been more favorable for harboring on the west side?
Answer. I have not the least doubt. I do not think one season is
exactly like the other. You will find that during one season you have a
good deal of wind, or low temperatures, and the consequence of it is that
the ice will freeze in hummocks.

Question. But you would probably be more helpless if you were cast
adrift and wrecked on the west side than on the east, there not being so
much assistance or so many Esquimaux settlements?
Answer. There are settlements south of Cape Isabella, and the natives
inform us that those Esquimaux have boats with which they could
actually cross the channel. They did actually cross with those boats
some years ago, and Jimmy and his family have remained on the Greenland coast; the others went back.

By Commodore REYNOLDS:

Question. By that time you would be at Smith's Sound; but the question is, as I understand it, if you were at Grinnell Land, whether you
would not find it more difficult to get away than if you were on the eastern side?

99

Answer. I do not think there is any point on Smith's Sound from where you could not reach the Danish settlement.

By Professor BAIRD:

Question. I mean still higher up—in latitude 81—away up as high as you can go on the west side. Would you do just as you did on this last occasion, go on the west side, or would you go on the east side?

Answer. As long as I had got ammunition to sustain life it would make no difference; there are plenty of musk-oxen there.

By the SECRETARY:

Question. Would not you find natives at as high latitudes on the west shore as on the east shore?

Answer. Certainly; that is just what I say. The natives on the west shore are, perhaps, a little more to the south; but, in fact, I do not suppose there is any danger if you are cast on shore at any point of Kennedy or Robeson Channel, if you have only dogs and sled and a gun.

Question. Was there any sickness on board except that of Captain Hall? Did you have to treat any one of the party?

Answer. We had a little scurvy among the crew, or a part of the crew, the first winter. The steward was taken sick with scurvy. It did not amount to much; he soon got over it. We had no trouble from any coughs or colds; in fact, nothing whatever. After we started for home Mr. Meyer had the scurvy a little, and during last winter some of our men had very light touches, but it never amounted to anything. We had not a sick-list during the cruise.

By the SECRETARY:

Question. Please give a description of your voyage in the Arctic.

Answer. After we were taken on board of the Arctic we crossed over in the Ravenscraig to the west side of Baffin's Bay. We went on board the Arctic near Cape Hay. We landed at Cape Hay to take some eggs on board from the resting-place, and, entering different inlets at Lancaster Sound, we took our way down Prince Regent's Inlet, landing at Fury Beach, examining the remains of the Fury that Parry lost at Fury Beach. Here we found scattered around some of the remains of the ship—a lot of the canned provisions in an almost perfect state of preservation. They consisted of preserved soups, meat, and vegetables.

Question. Why had not the Esquimaux got those?

Answer. Because there were none there. We tasted the provisions, and they were still very good. They were just as fresh as if they had been left there a short time since, and yet they had been there for more than fifty years. There was even some leaf-tobacco exposed to the air in barrels, and it had been wet several times. We took a few leaves along, dried and smoothed them, and they had not lost much of their flavor. Besides that, we found a cairn and thought it contained documents, and took it to pieces. But it proved to be a grave. I think most likely it was the grave of one of Ross's men.

Question. Had this place never been visited since Parry's time?

Answer. Yes, sir; when Ross had to abandon the Victory. He tried to get out of Lancaster Sound, but he had to come back. If I remember right he left one of his boats at Batty Bay, and had to put in there during the winter at Fury Beach. He built a house there.

Question. How do you know they were not his provisions?

Answer. Because he could not carry provisions along with him; and for another reason, that these provisions had the government mark on them. They had the broad arrow of the British navy. The Ross expedition was a private expedition.

Question. Had anybody else been there besides Ross before you went there?

Answer. As much as I know, Parry had been there, and Ross. In addition to these we found two English muskets, with the mark of 1850 on them. It is possible that some of the expeditions in search of Sir John Franklin may have visited this place, though I cannot now recollect that they ever did; perhaps Kennedy. We were on a boat expedition to the south shore of Creswell Bay, and, strange to say, we found thirty or more deserted huts of Esquimaux, built with the skulls of the Greenland whale. We found some ninety-six skulls. Captain Markham, with whom I made this journey on board of the Arctic, found a piece of rib belonging to a walrus that had been cut with a dull instrument.

Question. Are the specimens you brought home of great importance?

Answer. Yes, sir. Among the specimens that I brought back are some very valuable ones that will prove, among other things, that Greenland has been connected with America, and that a rupture took place in the direction from north to south. We found that certain minerals of South Greenland have been deposited as far north as latitude 82°. We did not find any of the Silurian limestone which composes all the rock about Polaris Bay and the newly discovered land south of Cape Constitution, showing that the drift was formerly from south to north instead of from north to south, as it is in these latitudes. Besides that, we know that North Greenland has been rising, because we found drift-wood and marine shells at elevations of 1,700 feet and more above the sea-level, shells that are found alive now in the adjoining sea. Besides that, marine animals have been found in fresh-water lakes, at an elevation of 38 feet. The land at some places rises in terraces, each terrace indicating one period of an upheaval. This is the land at Polaris Bay. It has corresponding formations in Prince Regent's Inlet and vicinity.

By Professor BAIRD:

Question. The west coast of Norway is rising, is it not?

Answer. Yes, sir; to a certain latitude, and then it is sinking.

Question. What do you infer from the fact that the tides of Polaris Bay seem to be connected with the Pacific Ocean rather than the Atlantic?

Answer. As a general rule you find that the night-tides along Greenland are very much higher than those that occur during the day; and there is hardly any difference at Polaris Bay. Besides that, we find that along the west coast of Spitzbergen high water occurs earlier the higher we get north. Consequently the said coast must be under the influence of two different tides.

Question. Do you infer from this fact that there is an open-sea connection between Robeson Channel and the Pacific?

Answer. Certainly. I have no doubt you can make a northwest passage if the ice does not obstruct you.

By Commodore REYNOLDS:

Question. The high water where Dr. Kane was occurred later than with you.

Answer. Certainly. Kane had the Atlantic tide, and his tide came from the south, while our tide came from the north. Kane's cotidal hour is later than ours. Our tide, as I say, came from the north, as proved by the establishment. We find by our observations that our tide came from the north, while Kane's tides, according to his account, came from the south. Ours came earlier than his, consequently ours could not have been the later effects of his tide, but must be an independent effect coming from the north. I therefore conclude at Polaris Bay the tides were the Pacific tides, not the Atlantic. The establishment at Polaris Bay, occurring earlier than at Rensselaer Harbor, proves that our tide must be a different one from that of Kane.

Examination of the witness being concluded, the commission adjourned until to-morrow morning at 11 o'clock.

WASHINGTON, *October* 18, 1873.

Examination of Dr. EMIL BESSELS resumed:

Question. Please take this book (Dr. Kane's Arctic Explorations, and the Second Grinnell Exploration) and examine the map in the fore part of it representing Kane's explorations, and state what corrections you are enabled to suggest as the result of your own observations.

Answer. In the first place, there exists another map—a second map—from the revised materials, in the Contributions of the Smithsonian Institution. That map has been constructed by means of observation and dead-reckoning, and in consequence of that we find that the most of the positions are too far north on this map. With regard to this map, beginning with Cape Constitution, we ought to place it, as I mentioned yesterday, in latitude 80° 25′ instead of 81°. Very likely the trend of Humboldt Glacier must be shifted a great deal toward the eastward. The northernmost point of the discoveries of Kane, on the east side of the channel, as laid down by astronomical observations, is Magarie's or Cache Island. The rest of the coast-line, in a certain direction, is correct, being based on a system of triangulation. In regard to the west coast, we have, in the first place, to mention that where Maury Bay, No. 25 on this map, is situated, a large sound ought to be shown as discovered by Dr. Hayes, and verified by the Esquimaux of Etau, who actually traveled through said sound. This sound proves Ellsmere Land to be an island. The sound itself is connected, very likely, with Jones's Sound.

Morton ascended an elevation of 500 feet, and it would be important to know how far he could actually see, because Mount Parry, put down as the northernmost peak seen by Morton, does not, in reality, exist, the land to the south of Lady Franklin Bay being of an entirely different character from that of the north. The former is mountainous, with a great number of peaks, like the coast of Spitzbergen, while the latter consists of a level, high plateau, with but a few hills. I tried to find the original survey of Kane, but I could not succeed. Parts of it are preserved at the Coast Survey, but other parts were at the Smithsonian Institution. Unfortunately, they were destroyed in the fire that took place at that building some years ago.

Besides that, the details of the west coast do not correspond to what in reality exists. The trend is also different, being more in an easterly and westerly direction. Instead of an open polar sea, as indicated by Kane, we found the land continuing, trending to the north and northeast up to latitude 81°, above Cape Constitution.

Question. Take Hayes's map and state what corrections you would suggest in regard to that. You have the chart of Hayes before you, as published by the Smithsonian Institution in January, 1865.

Answer. I have another chart, contained in the open polar sea narrative of Dr. I. I. Hayes, published in 1867. We find that there are discrepancies in these two charts. I refer to the plot produced the other day, and in looking at it we see that all Hayes seems to have done is to have shifted the coast-line, as laid down by Kane and Morton, a little to the westward, and making the different bays and inlets a little deeper. Comparing the chart in the narrative with that in the Smithsonian Institution, we find an island in the former in Carl Ritter Bay which is not laid down in the latter. Besides that, Lady Franklin Bay, on the former, terminates in a narrow inlet, abreast of which two islands are situated. The Smithsonian Institution map does not show the two islands nor the inlet alluded to.

In regard to the trend of Grinnell Land coast we noticed the same fact as stated when looking at the chart of Kane. The northernmost point reached by Hayes on his Arctic expedition is reached in 81° 31.5′, obtained by meridian altitude of the sun on the 17th of May, 1861. We all know that in such high latitudes the meridian altitude of the sun is not readily established, and perhaps the error of observation may sometimes be considerable.

In regard to the land north of Lady Franklin Bay I have to make the same statement I gave before; that is, that it is a high plateau and not mountains, as Hayes states it. Besides that, deep bays, like Peterman's Bay and Lincoln Bay, do not seem to exist—at least we were not able to perceive them from the Greenland coast right opposite. Cape Union could not be identified, and does not seem to project any above any of the other massive of the coast.

When we were at the farthest point the ship made, just a little southeast of Cape Union, we had the best possible chance that we could have to have seen it, but we could not see any deep indentation, and besides that we found the indentations much deeper on the chart in the narrative than they are in the Smithsonian chart. It seems to us that Lady Franklin Bay is a sound, on account of the setting in of the ice at a pretty brisk rate; and, standing on the summit of Cape Lupton, an elevation of 1,600 feet, we perceived a distinct separation of the north coast from something that seemed to be an island—a large island—in the middle of the bay. On Hayes's chart in the Smithsonian Institution it is marked Sylvia Mount, and on the narrative chart it is not marked at all. Mr. Meyer has given it on his chart as Mount Grinnell.

The little island on which we encamped on our sled at the northern point of the Southern Fjiord is in about the same latitude as Cape Constitution, as laid down on Kane's first chart.

Dr. Bessels here submitted a memorandum of the most important discoveries of the expedition, namely:

The results of the expedition may be summed up briefly as follows:

1. The Polaris reached 82° 16′ N., a higher latitude than has been attained by any other ship;

2. The navigability of Kennedy Channel has been proved beyond a doubt;

3. Upwards of 700 miles of coast-line have been discovered and surveyed ;

4. The insularity of Greenland has been proven ; and

5. Numerous observations have been made relating to astronomy, magnetism, force of gravity, ocean physics, meteorology, zoology, ethnology, botany, and geology, the records of which were kept in accordance with the instructions supplied by the National Academy, and some of the results of which we propose briefly to enumerate.

A₁—ASTRONOMY.

Great care was taken in determining a reliable meridian at Thank-God Harbor. Soon after entering winter-quarters an observatory was erected on the shore, thirty-four feet above mean sea level, and the transit instrument stationed there.

The longitude of this station was determined by the observation of—
300 lunar distances;
A number of moon culminations;
A great number of star transits;
A number of star occultations; and
A great number of altitudes of the sun on or near the prime vertical.
Its latitude, by the observation of—
A great number of circummeridian altitudes of the sun, and
A number of altitudes of stars.

All of these observations were lost, but a number of the results have been preserved which are sufficient to establish the position of this station.

The instruments used in the above observations were a Würdemann transit and Gambey sextants divided to 10″. The expedition carried six box chronometers made by Negus, three of which indicated sidereal time, and four pocket chronometers, by different English makers. These time-pieces were compared every day at precisely the same time, and the result entered in the chronometer-journal.

Besides the above-mentioned observations, 20 sets of pendulum experiments were made, which are saved, but the observations for time belonging to them are lost.

B.—MAGNETISM.

The magnetic observations obtained were more complete than any others ever before made in the arctic regions. The instruments supplied were:
One unifilar declinometer;
One dip circle, with Lloyd's needles;
One theodolite; and
Several prismatic compasses.

The observations on variation of declination were registered at Göttingen time, and were continued for five months. Readings taken hourly. Besides that, three term days were observed every month, according to the Göttingen regulations, one of these term days corresponding with the day accepted by all the magnetic stations. Further, a number of observations were taken either with the theodolite or the prismatic compass. Whenever possible, the dip was observed, and several sets of observations on relative and absolute intensity and of the moment of inertia were obtained.

C.—OCEAN PHYSICS.

Unfortunately there was not much opportunity for taking soundings. About 12 were obtained along the coast of Grinnell Land, which prove that the hundred-fathom line follows the coast at a distance of about 15 miles in Smith Sound. One of these soundings (90 fathoms) proved highly interesting, containing an organism of lower type than the *Bathybius* discovered by the English dredging expedition. It was named *Protobathybius robesonii*.

A number of deep-sea temperatures were taken with corresponding observations on the density of the water. Following the coast of West Greenland the limits of the Gulf Stream were ascertained. Specimens of water from different depths were preserved in bottles, but, unfortunately, lost.

As soon as the vessel was fairly frozen in a tide-gauge was erected over a square hole cut in the ice-floe, and kept open continually; the pully and rope were supported by a tripod of oars. A rope to which a wooden scale, divided into feet and inches, was fastened, was carried through a block attached to the tripod. One end of the rope was anchored at the bottom by means of two thirty-two pound shot, and a counterpoise was attached to the other end to keep the rope properly stretched. This apparatus was tested by a series of scale readings with corresponding soundings, and proved to work very satisfactorily. The observations comprise eight lunations, the readings being taken hourly, half-hourly, and in some instances every ten minutes, in order to determine the precise moment of the turn of the tide.

D.—METEOROLOGY.

After having entered winter-quarters meteorological observations, which up to this time had been made three-hourly; were made every hour, Washington time. The register contained observations on the temperature of the air, atmospheric pressure, psychrometrical observation, direction and force of wind, appearance of the sky, state of weather, and both solar and terrestrial radiation. Besides, all extraordinary meteorological phenomena were carefully noted.

For the registration of the temperature of the air mercurial thermometers were used for temperatures down to —35 °F.; for lower ranges spirit instruments being compared at intervals of 10 degrees. As circumstances would permit, mercurial or aneroid barometers were used. As it was not supposed that psychrometrical observations could be favorably conducted at very low temperatures, the expedition was not supplied with the suitable instruments. For that reason two uncolored spirit thermometers were selected and used, the readings of which agreed. As check observations the dew-point was determined by means of Regnault's apparatus. To measure the velocity of the wind, Robinson's anemometer usually served. The distance traveled by the wind was noted hourly, at the same intervals of time. The velocity of the wind was determined either by the same instrument or by means of Casella's current-meter. These observations on the winds, combined with those on moisture of the atmosphere, will form a valuable contribution to physical geography.

It was not thought essential to procure photographs of the clouds, as they do not differ in their general character from those in more southerly latitudes. The only remarkable fact to be noticed is that sometimes cirri could be observed at very low altitudes among stratus clouds, which, however, is not surprising if their mode of formation is taken into account.

Special attention was devoted to the aurora borealis, which occurred frequently, but rarely showed brilliant colors, never bright enough to produce a spectrum. Whenever necessary one observer was stationed at the magnetometer and the other out doors, the former observing the motions of the magnet, while the other was watching the changes in the phenomenon and taking sketches. Although an electroscope and electrometer were set up, and the electrical condition of the atmosphere frequently tested. In no instance could the least amount of electricity be detected. The amount of precipitation was measured as carefully as the violent gales would permit, by means of a rain-gauge supplied with a funnel. In February, as soon as the sun re-appeared, observations on solar radiation were commenced, and continued throughout the entire summer. The instruments employed were a common black-bulb thermometer, and one *in vacuo;* both exposed on white cotton.

E.—ZOOLOGY AND BOTANY.

The collections of natural history are nearly entirely lost. With the exception of two small cases containg animals, minerals, and one pack age of plants, nothing could be rescued. The character of the fauna is North American, as indicated by the occurrence of the lemming and the musk-ox. Nine species of mammals were found, four of which are seals. The birds are represented by twenty-one species. The number of species of insects is about fifteen, viz: one beetle, four butterflies, six diptera, one bumble-bee, and several ichneumons, parasites in caterpillars. Further, two species of spiders and several mites were found. The animals of lower grade are not ready yet for examination.

The flora is richer than could be expected, as not less than seventeen phaneragamic plants were collected, besides three mosses, three lichens, and five fresh-water algæ.

F.—GEOLOGY.

Although the formation of the Upper Silurian limestone, which seems to constitute the whole west coast north of Humboldt Glacier, is very uniform, some highly interesting and important observations have been made. It was found that the land is rising, as indicated, for instance, by the occurrence of marine animals in a fresh-water lake more than 30 feet above the sea-level and far out of reach of the spring-tides. Wherever the locality was favorable the land is covered by drift, sometimes containing very characteristic lithological specimens, the identification of which with rocks in South Greenland was a very easily accomplished task. For instance, garnets of unusually large size were found in latitude 81° 30', having marked mineralogical characteristics by which the identity with some garnets from Fiskenaes was established. Drawing a conclusion from such observations it became evident that the main line of the drift, indicating the direction of its motion, runs from south to north.

It would lead too far to enter into detail with regard to numerous miscellaneous observations that were made besides those mentioned above.

Examination of Emil Schuman.

I was born in Dresden, capital of Saxony. My profession is that of an engineer of bridge and road building, and laying out streets, &c. I am thirty years of age. I joined the Polaris expedition in Washington before it started. I joined it here as chief engineer of the steam-department. I went on the Polaris to New York, New London, and Greenland, and arrived at Disco and lay there until the Congress came. We took provisions in there, and also coal, and started for Upernavik. Nothing that I know of of any note happened at Disco. At Upernavik we took dogs, seal-skin, &c., and then started for the north. From Upernavik we started up past Fitzclarence Rock. I made a drawing of that at the time. We found near Fitzclarence Rock great quantities of pack-ice. The water, however, got clear pretty soon again, and we worked through with steam. We used steam all the while, not having our sails up at all. We steamed up in the ice as high as 82° 16′. That is as high as it is made by the last correct observation. Captain Hall thought at first that we had been still farther up. He said we had reached as high as 82° 26′. When we got to that point we saw that there was no chance to get any farther; we therefore made fast to an ice floe and came back. I think it was the 1st of September when, an ice-floe coming against us, the vessel got a nip. Captain Hall thought we were in great danger, and ordered the provisions to be taken onto the ice. All hands were set to work at this duty. He told me to make a sketch of the position of the Polaris, and I did so. That sketch has been preserved. As soon as the danger passed away we brought the provisions on board again. We found that we had drifted down a good deal to the south by the current. Then we came out of the ice into open water again. That, I think, was in the night, from the 3d to the 4th of September. We steamed northeast and reached 81° 38′, where we went into winter-quarters. Captain Hall · then gave an order to keep steam up all the time until we were frozen in, in case something should happen. Then the next day he thought he could find another harbor farther south, and told me to get steam up. We steamed south, but did not succeed. We then came back again to the same place. I think five days after that he told me not to fire up any more. Then I went to work and took the engine and everything apart, all the pipes, &c., so they would not freeze up. Expansion might have caused them to burst. The next thing that occurred was Captain Hall's going off on a sled journey. I really forget when it was that he started. I had everything written down and then lost the memorandum. I know that he was gone fourteen days. He had divine service every Sunday, and one day he told us that he was going musk-ox hunting. Dr. Bessels and Mr. Chester had been on such an excursion, and had brought one back, and he thought there must be plenty of musk-oxen in the country there, and so he gave it out that he was going musk-ox hunting for the purpose of bringing in fresh meat for the winter time. That is the only thing that I knew of. I knew of no other purpose. He then went on this musk-ox hunting excursion. He started with one sled, but afterward sent back for another. I think he took with him some fourteen dogs. Mr. Chester and the two Esquimaux, Joe and Hans, accompanied him. He was over fourteen days away, I am sure. He came back at the expiration of that time, bringing nothing with him. He said he could find nothing. When he came back I met him on the vessel just as he came on board, and I asked him how he felt. He said "pretty well." Then he went into his cabin and I went into my room. In the evening when I came into the cabin I found him sick in bed. I had at that time myself a very bad cough and remained on board of the ship most of the time. - The next day I asked Dr. Bessels if he could give me something for my cough. He said that I should remain in the room. I went into the cabin where Captain Hall was, and I was there in the cabin with Captain Hall during the whole fourteen days he was sick; I only went out when I had occasion to. Being in there so much of the time I heard everything that was said, except when I was asleep, of course. In a short time he got delirious and remained so for the first three days. He really did not know what he was doing or saying. The fourth day his head was clearer, and I thought he was getting better. He could speak and we thought he was all right. Then he laid down in his bed and jumped up and got crazy again. He would take his book and commence to write and then walk around in the cabin and suddenly change right off again. It was not three days after that before he was dead. He got very sick that night, and I believe on the 8th of November he died. On the evening of the 7th of November I heard him call the doctor and say to him, "Doctor, I am very much obliged to you for your kindness," and in the morning at 3 o'clock, I think it was, of the 8th, he was dead. He never said a word more than what I have just alluded to as having been said to the doctor that I ever heard after that. He laid perfectly quiet—could not move his left side three days before his death. I do not think that he had paralysis when he was first taken sick. I observed that he was better, and then I saw that he could not move the left arm at all, and when he walked in the cabin, after he got better, that was hanging down all the time. I do not think he had any difficulty in moving his left leg, but I always saw that arm hanging down. When he got into bed he would take hold himself of his left arm with his right and lift it up. I never noticed him in any kind of stupor. I have seen him sleep heavy, and he seemed, at times, to sleep very well. He would ask sometimes what that blue thing was coming out of the mouth of some person, and then he would call different people at times, and when they came he would call for some one else. Most of the time the doctor was in the room. He accused Mr. Chester of trying to shoot him. He would say to him, "I am not afraid of your powder." At one time he sprang out of bed and grabbed hold of him. Captain Tyson and Captain Buddington seized Captain Hall and put him into bed again. Whenever they heard a noise in the cabin they would come in to see what the trouble was. The cook was, at first, always by his side, but Captain Hall made him go away; he said he did not want him in the cabin with him any more. He thought he was going to kill him. I never heard him speak about poison in any connection; but everything he would eat he would first make us taste it. I never, however, as I say, heard him speak of poison.

I used to taste his food myself. He had a certain kind of beans that he sometimes ate. I do not know the name of them. I tried that, also. Everything that he ate somebody had, in the first place, to eat of. I never heard him say what was his reason. I am sure I never heard him speak about poison. When he thought he had offended any one he would, after a while, beg his pardon. He begged my pardon about ten times. He used to say to me, "Mr. Schuman, if I ever did wrong to you, I beg your pardon; I am extremely sorry." He said this to most every one. At one time he called Captain Buddington and told him in case he should die that he, Captain Buddington, should go to the north pole and not come back before he had reached it. Captain Buddington had to promise him that he would do so. That was about five or six days before his death. He was then a little better. That was the only time I heard him say anything about going to die. He said he would not live until the next day, but he lived about six days afterward. He did not say anything in my presence as to what he thought was the matter with him. He did not seem to notice his paralysis. He did not talk much about it. Hannah was sometimes in his room, and Joe and Hans sometimes came in. Dr. Bessels was with him most of the time. All those in the cabin, Mr. Meyer, the cook, the steward, and myself, were with him a greater part of the time. From the fact of my suffering with the cold I have spoken of, I was in the cabin myself nearly all the time and saw nearly everything that happened. I had a very severe cold and not able at times to speak. I will state that Captain Hall was kindly taken care of by every one. The doctor was especially kind to him and did everything he could. The doctor, for instance, had a string on his arm and he made that fast to the arm of Captain Hall, so that, in case the captain wished anything, he had only to pull the string and that would notify the doctor. There were very few hours, indeed, that the doctor had sleep. The string was frequently pulled by Captain Hall. He seemed to want him all the time. He would not take medicine, however. I saw the doctor attempt to give him medicine, but he would not take it. Captain Buddington, Mr. Chester, and all of them begged him to take medicine in order that he might get better. He did take a little. He could not eat much. I did not see him eat things that the doctor did not want him to eat. He never took any great notice of the doctor in that respect. I did not see him himself open any canned meats to eat; I heard it stated that he did so, but I do not think he did. I used to open some canned meats for him, but he would not eat it; he gave it away again. He used to ask for everything. We used to indulge him by preparing such things for him as he expressed a wish for, knowing that he would not eat them when they were given to him.

I believed then, and believe now, that Captain Hall died a natural death. I saw my father die just in the same way that he did. I knew right off when Captain Hall was dying. I did not have then, and have not now, any suspicion as regards Captain Hall not having died a natural death. I do not think any person on board the vessel had. I never heard it intimated that he died from any other than a natural cause until I got to Dundee, and then I saw some such intimations in the papers. After he died we made a coffin and buried him. His journal and papers were all saved, I think. The captain took them in charge; he put them in a tin box, and read them, and we all read them. I did not read them myself, because I could not read Captain Hall's hand-writing. I tried to do so, but could not. His journal was kept in a book like this, (referring to one of the books upon the table.) There was not much writing in the book. He commenced to write in it when we were in winter-quarters. He never did anything before that. Mr. Meyer kept his journal before that time.

Captain Buddington took command after Captain Hall's death, and Dr. Bessels took charge of the sledge-journeys and the scientific observations. The discipline of the ship was very good with both Captain Hall and Captain Buddington. It was just the same with Captain Buddington as it was with Captain Hall, only we had a little more liberty than we had when Captain Hall was in command. Captain Buddington told me that he had to give this liberty to the men in order to prevent the men from getting sick. He seemed to regard it as necessary that the men should have more or less freedom. He contended that that was the only way to keep sickness away from the men. The discipline, however, was well preserved all the while. I never heard one word out of the way. There was no disorder in the ship after Captain Hall's death, that I am aware of. There may have been forward in the forecastle, but I did not know of it. As regards Captain Buddington's habits of drinking I will say that Captain Buddington was tipsy sometimes, but I saw Captain Tyson drunk like the old mischief. I saw Captain Tyson when he could scarcely move along. We were in winter-quarters at the time. That was after Captain Hall's death. After a while there was nothing more to drink. I think there was only about one hundred bottles of whisky on board. There was no general drunkenness on board at all. Captain Buddington was drunk, I think, once or twice. We remained in our winter-quarters until the 12th of August. During that winter I wrote a journal, but I lost it. I made the drawings, and during that time copied them off. I had put the machinery back again in the spring. When the temperature was warm enough to admit of it, I connected every one of the parts. There was some little repairing to be done. During the spring we found that the ship was in a leaky condition. I told the captain about it. I said to him that there must be a leak in the vessel. He said, "O, no; it is the water running off from the melting snow." I told him that I did not think that would make as much water as there was. We then commenced to examine the ship, and Captain Buddington found the leak outside. It was forward at the stem, 6 feet. The stem was broken. I do not know what could have been the cause of it, but I know that on the 21st of October, at the time we had a northeast gale, we broke out of winter-quarters and went adrift. As soon as we saw that we were adrift, and noticed that there was danger for us all, Captain Buddington ordered the second anchor out. We threw it out, and it got aground, and the vessel swung around onto the iceberg; there we made fast. One of the sailors went out and fastened a hook into the berg—we had fearful weather—that is all that saved us. The next day the weather got better, and we saw that we had all around us open water. The third day after that, when we were frozen in, we could walk on the ice again. Captain Buddington then ordered us to saw the vessel out from the iceberg. I think it was about two hundred yards. On the 27th there came up a southwest gale, and the gale hove the iceberg against the vessel. The tongue of the iceberg went underneath the vessel and struck the stem of the ship, and that wrenched it.

During the whole winter time the ship rested forward on that tongue, and aft she was afloat in the water, and then she was moved up and down by each tide. That broke the stem. We tried everything we could to stop the leak. We endeavored to calk up the place, and took off the iron plates and nailed them fast again. We could do this on the starboard side, but not on the port side. She was too much in the water there. I thought the leak was stopped until we got afloat again. As soon as we found ourselves in open water we discovered that the vessel was leaking. During the time when the party were to the north in those two boats—Mr. Chester and Mr. Tyson, with their different crews—we had to pump the vessel by steam. When they came back we pumped the vessel by hand with the large pump. We had made three different attempts to go north in August and before August, before we started for home, but we never succeeded in getting north. When we started for the south I pumped the vessel with the big engine because I had steam. When we made fast to the floes we were beset so that we could not go on. Then we waited for a chance to go on. I pumped only with the steam I had left. When that was all gone then we pumped by hand. I thought it, then, a very good thing for us that we had that little boiler that we intended to use for burning blubber. I took it and brought it more aft, and connected it with the feed-pump, so as to use the little boiler to pump the ship out. I succeeded in making such connection, and kept her just about clear. She was steady, going without stopping, for as many revolutions as she could make with the little pump.

One day we got a nip by a floe, and I found that she was leaking more. I could not pump any more with that little pump, and I set a hand-pump on again, and pumped with that. The sailors did the pumping and kept her clear until that night, when she broke adrift. I think it was the 15th. Before that we had been building a house on the ice, so as to provide a shelter in the event of our having to leave the ship. We had all the provisions in readiness and had the coal in bags and ready to throw over when the time came. We did not throw off all the coal there. We had ten tons more in the ship.

I had all the while a temperature in the engine-room, which was above freezing-point all the time; but I was always able to get steam, and had a little boiler full of water constantly. When we found that we could not keep afloat by using the hand-pump, and I saw the water gaining, I reported to Captain Buddington the fact. I said she would not last more than three hours, and would then sink. I said "We cannot make steam then, if we cannot now." I saw the water gaining in five minutes about two inches, and the pump all the while going. As soon as Captain Buddington heard my statement he said, "Shove everything overboard." I made steam without any order. I saw it was the only chance to do so. I saw that, owing to the confusion, there was no probability of getting any order to that effect, and I took the responsibility upon myself. Before I had steam up, however, we were adrift. I came up and saw that the vessel had broken off, and saw some of the people on the ice. Perhaps in half an hour after that I had steam up, and the steam-pump gained an inch in an hour. I had to use the little steam-pump. We had no coal to do anything else with. We put six tons of coal on shore before we concluded to pump the water. There were ten altogether. If I had used the big pump, all the coal would have been used up in four hours. I never made a report that the water did not gain. That is what I have heard, but I never said such a thing. I was the only person who had an opportunity to know about this matter. The others were afraid to go below. The next morning we saw land; as soon as it was a little daylight, we saw land, and we looked for a chance to get on shore. We sent Mr. Chester up to the mast-head to see if he could discover anything of the other party. He reported he could not see anything. I had been up myself, but I could not see anything there. I had been up there only once. I could not leave my post at the engine a sufficient length of time to be going up there. I had a good look around while I was up there, however. The weather was clear. I had a glass with me and yet I saw nothing of them, nor of the house on the ice. I saw nothing but the water around. We steamed in-shore. The little boiler gave me about 600 or 700 revolutions as we worked steam up. Then I waited a little while until I got steam again to make those revolutions again, and so we worked ourselves through the ice and came on shore by high water. At low water we were aground. Then we went to examine the vessel, where she was and the condition she was in, and we found that the whole bow was gone. The six-foot piece was broken away entirely. The water-tight compartment, which the carpenter had built up in the winter, was the only thing that saved us. The ends of the plank were gone where the planks joined on to the stem. You could see into the bow of the ship. I made a sketch of it, but I cannot find it now; I do not know what has become of it. The ship's bow was open so wide underneath that you could see in and almost enter the boat through it. A man could have crawled in there, I am sure. She was beyond the reach of repairs that were within our means at that place. We might have repaired her in a dry-dock. I made up my mind that we would have to abandon the ship; that the vessel would have to stay there; that we would have to build boats. I did that as soon as I saw the condition that she was in. Then we went to work to build a house, and prepared to pass the winter there. This was in the middle of October, I think, of last year. We all enjoyed the best of health during the winter. Nobody was sick at all. We had 10 boxes of meat, and had plenty of dried apples, sweet potatoes, Irish potatoes, and then we had fine seals brought on shore, and we had plenty of bread.

Captain Buddington during all this time that we were fast to the ice-floe, when we separated from it, and up to the time we ran ashore, did everything to preserve the Polaris from wreck. He could not do anything more. It was not possible. What he could do, he did. He was very anxious. He very often asked how it was, &c., &c., but you see I was mostly at the engine. I could not hear much that was going on on deck, but I know that he was a gentleman that did everything he could do as captain. I would cheerfully go with him again to the Arctic region if he were to go. My engine is all right yet I think. The only trouble is she is in the water. I had no trouble at all with her. The steam department worked well during the whole cruise up to the time she was beached. She was well fitted in the steam department.

In the spring we began to make boats. Mr. Chester and the carpenter built three boats. We gave one to the Esquimaux—the little one—and the others we kept. The boats were not coppered on the bottom, but they were very good—very well built.

On the 3d of June we started. We had a fair wind. We embarked in the two boats for the purpose of making our way to the South Greenland coast. We worked down about three hundred miles with fair weather. We never in fact had bad weather at sea. On shore we had several storms. When we saw a storm about to come up, we would put into shore and remain until it was over. Thus we continued until we were picked up by this steamer, below Cape York. It was very fortunate that we met with her at the time we did, as we had no fuel on the boats to make a fire to keep us warm. We had used the last piece just that day when we saw the Ravenscraig. We burned up all the coal at winter quarters. The coal was gone in January. That was all the coal we had on deck. We never used the Disco coal. It is there still. We had no chance to go down to get it. The coal we had then was from Washington. I saved the best coal until the last, and that coal was the coal we got from Washington. We saved enough wood from the vessel with which to make our boats. Then we commenced to take wood from the vessel for the purpose of fuel. When we left, all that could be seen of the Polaris was her deck. The rest was under water, and so she was the whole winter through.

I know of no difficulty at Disco or elsewhere. I did not see anything, and I know nothing except what I have heard others say. I really had nothing to do with any one else on the boat. I was kept pretty busy attending to my engine. It was only in the winter time that I went into the cabin. I used to go in there and remain there because I had no fire in my room, and therefore could not remain there, of course. My machinery would have been injured if I had not taken it all apart. I took the brass from the iron because it does not expand equally. I took everything apart and laid it one side. I think the excessive cold would have disabled the machinery if I had not done this. That will occur in the winter, almost everywhere, if care is not exercised. I think it was less likely to rust in that region than elsewhere, because the weather is drier. I did not of my own knowledge know of any difficulty that occurred between Captain Hall and anybody during the cruise. I am so constituted that I would not hear any if it were to take place. I would go away. As far as the crew are concerned, they were very obedient to the officers. There were difficulties sometimes, but nothing serious. The only difficulty that occurred between the officers and men was because the men would not sometimes observe discipline; but Captain Buddington would generally bring everything into good order by a few kind words. I have never been on board of a vessel where there was so much harmony as on board of that vessel. I have now been at sea twelve years. I have always been an engineer; not on a Government vessel, however. In the first place, when I came from the polytechnic school, I got a position in Marseilles in a manufactory there, and from there I started off again to put an engine up in Africa, in Philipsville. Then I stopped in Africa three years and engaged in road-building, streets, &c. I worked at that three years, in different towns, such as they sent me. I built a market once. From Africa I started over to France again, and came home. I had the fever and ague at that time. As soon as I got well I came to America, and got a place in the North German line as assistant engineer. I worked myself up to the position of an engineer, and then I stopped in America. Then I took a place with Captain Hall. I was three years with the North German line. I gave up a good place to go on this expedition. I think the discipline on board the Polaris was not as good as that on board the North German line. We had a little more liberty on the Polaris than on the German line. Those liberties could be afforded, because we were free from temptations on shore such as are encountered on the North German line and other lines of steamers.

Examination of Henry Hobby.

I am a seaman. I was born in Germany. I have been to sea seventeen years. I have sailed in merchant-vessels as a sailor seven years; as first mate, two years; and as captain, I have been all over the world pretty much. I sailed as captain in the Mediterranean and North Seas, East Baltic, &c. They were small ships. I have been in merchantmen, but never in ship of war. I was first mate in an American bark that belonged at New York. She went to Callao. I was in no other American ship. That was in 1869 and 1870. I did not come back in the ship. She was condemned in Callao, and she had no cargo in her. I went ashore, and returned to Hamburg. I joined the Polaris at New York. Mr. Morrison engaged me as a seaman. Then I went in her from New York to New London, and from there to St. John; thence to Greenland, and thence to Disco. Nothing of importance occurred on the way. From Disco I went up north to Upernavik. There we got Hans and his family on board; and went from there to Tessiusak. We left Tessiusak on the 24th of August; we steamed north, always along pretty well through slack ice. On the 27th of August, we passed Hayes's winter-quarters at 3 o'clock in the afternoon; the next morning at 7 o'clock, Captain Hall landed on the west coast, off Cape Frazier, looking for winter-quarters, but could not find any. The ice opened again, and we steamed farther northward. In the evening we passed Cape Constitution, and we saw land on both sides off Cape Constitution, and after passing it. We were told, after passing Cape Constitution, that we would be in the open polar sea the next morning. Next morning, however, we continued to see land on both sides of us. The weather was not very clear; a little foggy; but we could see the land on both sides, notwithstanding. It is a very narrow channel. We steamed up the next day for the north. The next morning, I believe it was the 28th of August, we got stopped in the ice. We turned back at 6 o'clock in the morning. We continued still to see land on both sides of us. On the 29th, Captain Hall called all the officers on the house, for the purpose of having them consult as to what it was best to do about establishing winter-quarters, or going farther north. I was on the lookout at that time on the crow's-nest. From what I heard, nearly all the officers wanted to go north. Captain Buddington and Captain Tyson said it was necessary to make winter-quarters as fast as possible. I could hear every word that was uttered. Captain Buddington wanted to go into Newman's Bay ; Captain Hall and all the rest wanted to go north, with the exception of Captain Tyson. That is what I think. I heard no expression from Captain Hall; he merely asked every one where he wanted to go. I did not hear Captain Hall say himself what he wanted to do. When I was up there in this crow's-nest, and they were talking about it, I could see a way for going north on the eastern shore, from north to about northeast. So far as I could observe, I saw open water. There was land on both sides. There was no ice between us and the open water that I saw. I sung out from the crow's-nest, inquiring where they wanted to go. I told them there was plenty of open water to the 'northeast. I could not see exactly the point. Captain Buddington said that we must make winter-quarters. These were just the very words he said. I asked him where he wanted me to go, and he said, "Right over there, to Newman's Bay." The ship was lying still at this time, under steam, and not fast; she was just lying there. There was no ice to stop us from going north, as far as I could see. We steamed across towards the west side. We were about in the middle of the straits when we got beset in the ice. It was eight or ten hours after we commenced to steam west before we got into the pack. I was not in the crow's-nest all that time. Captain Hall called all the officers up there at 12 o'clock, or a little after 12. I came down by 2 o'clock. The ship was steaming among the ice when I came down, crossing the straits. We got, as I say, fast in the pack. The gale commenced from the northeast; we drove down with it, and when it came down Captain Hall or Mr. Chester crossed that floe where we were lying to see if he could not ascertain whether there was any passage over to the land. They went on the Greenland side. They went on foot. There was a floe of about five miles. When he came back Captain Hall ordered all provisions on board. In the morning, between 7 or 8 o'clock, we took everything on the vessel. We had put out a lot of provisions during the night. We put them out on the first of September, but took them back on the 2d or 3d. It was daylight all the time we were putting them out, but the sun was not shining. After we took the provisions on board, Captain Hall ordered steam to be gotten up. At 9 o'clock in the evening we opened a little bit of lead into the Greenland coast ; that was about three or four miles from the coast. We had only the little boiler working. At 12 o'clock we dropped our anchor in Polaris Bay. The next day we laid there. In the afternoon we steamed down south looking for a better harbor than that was, but could not find any other place, and therefore made that for winter-quarters, up behind Providence Iceberg. At that time there was plenty of open water in the straits, and we were going to run out of it. It was calm weather, and no frost. We could see plenty of open water from the deck of the ship. Some of the officers wanted to go north, but some of them wanted to stop there. I heard them say so. Mr. Chester was one who wanted to go north. There was no one else that I heard say that. Captain Buddington said

8 P

it was the safest place to make the winter-quarters; that the season was too late to try any more to get north. I thought at that time that it would have been best to have steamed up to Newman's Bay to make winter-quarters there. We had seen that before. There was no ice coming down with the northerly wind in the straits. It was only twenty-two miles from Polaris Bay. I could not say whether we could have got any further than Newman's Bay or not. We made the ship secure there and commenced to sledge provisions on the shore. That is, we fastened two boats together and made a bridge over that, and took the provisions on to the shore, and built a house out of provision boxes. It was afterward broken down. I mean boxes that had provisions in them. We used the boxes for making the side-walls, and were going to put a sail over them, to have a house for exercising and amusing ourselves in. Mr. Chester and the carpenter were making a floor to it, &c. We intended to have everything nice and comfortable, but afterward it was left. Provisions were stored all in one pile. Captain Hall ordered the house taken down before he went away on the sledge-journey. He started away on the sledge journey, I think, on the 14th October. He was away a fortnight, I know. Hans, Joe, and Mr. Chester accompanied him. They had two sledges. He came back a fortnight afterward, at half past 1 in the afternoon. We, all hands, were outside banking up the ship with snow. We made a snow-wall around it. I was attending to the tide observations at that moment. Captain Hall came to every one of us and shook hands, telling us how far he had been. He looked first-rate. A little while afterward, in about an hour, we were told that Captain Hall was sick. We heard it from the steward and from the captain. We were forward then on the ship. The steward told us that Captain Hall did not feel well. He said he had turned in, after he had drank a cup of coffee, and that Mr. Morton had undressed him; that he did not seem to feel quite well. Next Sunday, after Captain Hall had somewhat recovered from his sickness, the Esquimaux shot a big seal. Captain Hall was pretty well that day, and walking up and down the cabin. He seemed glad at the fact that a seal had been killed. I saw him myself at that time. I had not seen him before after his sickness. We were not allowed to come aft. I helped to carry the seal on board. It weighed five hundred pounds. We carried the seal-meat aft of the wheel-house, for the purpose of having it there if any of the men got sick with the scurvy during the winter. He seemed to be rejoiced to think that he had it. He did not so express himself to me, but I heard him say to Hans that he was so glad that he had got a seal. I could see him through the windows, and I saw him laughing and rejoicing over the fact. Three days after that he was dead. I did not see him after he was taken sick until this once, when I saw him through the windows, when I was carrying the seal aft.

Joe Mauch, captain's clerk, came into the cabin in the morning and told us and told the chief engineer and myself, that there had been some poisoning around there. I asked Mauch about it, and he told me that there was "Ilousaure." I do not know what it is in English. He did not say any more about it. I do not know what it was used for, whether it was good or bad. He did not mean to say that Captain Hall had taken this, but that the smell was in the cabin—used there for some purpose or other. Captain Hall died at half past 3 in the morning. Some of us were called out. I was awake, and was told that Captain Hall was dead; Captain Buddington told me. I came up on the deck; he stood just on the fire-room scuttle, and said, "We are all right now." I said, "How do you mean by that." He says, "You shan't be starved to death now, I can tell you that." I says, "I never believed I would." That is what I told him. I had not been starved before. We had been pretty well hungered, however. Half of us only got enough to eat. This was without Captain Hall knowing it, for Captain Buddington told us here in Washington, at the navy yard, that in regard to all matters of eating we had to come to him, and not go to Captain Hall at all. We never went to Captain Hall at all. We always told Captain Buddington that we had but half enough to eat. On one occasion two of the men were sitting down in the lower cabin, Captain Hall was in the upper cabin. He was making a tent that he intended to use on his journey north—a canvas tent. Mr. Chester says to me, "Well, boys, it is dinner time, and you can go and have your dinner." We said, "All right, but it is not much to us; by the time we come to eat after 12 o'clock, it will all be gone any way." Captain Hall heard the remark from where he was in the upper cabin. This was before Captain Hall started on his sledge journey. The next morning and every morning Captain Hall was alive, at half past 8 o'clock in the morning, we had to be up, washed, and dressed, and clean, and in the upper cabin for prayer. Everybody had to be there. Just at the close of the next meeting, after he had heard the remark I have referred to, he told us that he had taken the expedition from the American Government, and we all must eat and drink alike. That we were servants to him, that everything he had to eat and drink we would have just the same. Captain Hall said that he did not know that we were not fed sufficiently. This conversation was in the upper cabin in the morning at prayer-meeting, in Polaris Bay, before he left on the sledge journey. All hands were present when he said this—officers and men. John Heron, the steward, served out the provisions. We did not have enough to eat after we left Disco ; there was some shortening before we got there. There was some kind of quarreling, that is, we had heard that there was some kind of mutiny on board; when in fact there was no such thing. We heard that Captain Buddington, Dr. Bessels, and Mr. Schuman, and all of them were going to leave, and there was some talk among the men about leaving also. This was at Disco. What we said was, that we did not have exactly enough to eat, but then it did not amount to anything at that time. We never spoke to any of the officers, saying we were going to leave. I did not intend to leave, but some of the men said that they were going to leave; they were afraid that Captain Buddington, Dr. Bessels, and the others would leave the vessel, and that we would have all the regular naval officers come aboard from the Congress, and that they did not seem to like.

From St. Johns to Disco we had plenty to eat, then there was a change of rations at Disco and afterward. It commenced as soon as we got into the ports of Greenland, Fiscanaes, Holsteinsberg, &c. It changed slowly. The rations changed in every respect. At dinner time, if we did not look out and get there in time, there were three or four who would get there and take everything on the table and the rest would get scarcely anything. That was not in Disco, but that was afterward. In Disco it commenced to shorten down all the while. It was not because we were able to eat so much more when we got up into the northern latitude ; there was not as much served out to us when we got up there. We spoke about it to Captain Buddington and he said he would look after it. We never got the thing made better until Captain Hall found it out himself. I do not know why the allowance was shortened. Captain Buddington told us at the Washington navy yard that we would have plenty to eat. After Captain Hall found it out we then had abundance—as much as we wanted. The cook generally spoiled the grub, however. I was forecastle steward. Captain Hall allowed one man to look after the forecastle for the purpose of keeping everything clean with a view to health and so on. While in the performance of such duty I heard Captain Hall saying to the cook myself, that if you are not down attending to your business you shall not have a cent of pay when you get home. He said that on the galley while I was standing alongside of the galley door. The cook was a mulatto man. The ship was then in Polaris Bay; I could not say whether it was exactly a couple of days before the prayer-meeting to which I have referred, at which Captain Hall made the remarks relative to our being better supplied or not. The cook I know never did any better. He spoiled more grub than there was on board the Polaris, and if there had been any more he would have spoiled more. He never cooked a proper meal. There was plenty of everything. The bread, however, was half baked; so was the musk-ox meat. His bread not being in a condition to be at all eatable he would throw the dough over the side.

We were always pretty short at sea from Disco on to Polaris Bay. We could eat pretty well what he gave us. We had meat, potatoes; but the potatoes were not at all boiled. We had only one barrel of salt beef with us on board. Captain Buddington came out after Captain Hall's death and said we would have plenty to eat after this. Nothing further took place then. I went away into the forecastle. I did not know what he meant. I went down in the forecastle and cleaned it up, when one of our men was helping the carpenter to make a coffin. Two days after that Captain Hall was buried. I think it was on the 10th of November. After he was buried, everything went along as Captain Hall ordered it, in every respect, only the prayer-meetings in the morning; they were discontinued shortly afterward. Captain Buddington discontinued them. All hands, officers and men, were at prayer-meeting one morning, and he told us, when we got through, that the prayer-meetings would not be continued any longer, and that every man might do his own praying there. Mr. Bryan conducted the prayer-meetings for some time after Captain Hall's death. We had prayer-meetings on Sunday from 11 to 12 o'clock. Everybody had to be there while Captain Hall was living, but about a month or six weeks after his death they were discontinued. There was only Herman Siemons, Captain Buddington, Mr. Morton, Hannah, and myself in the prayer-meeting one Sunday morning, and Captain Buddington said there was no use of carrying this on any further; nobody wants to come any more, and we had better knock it off altogether. Mr. Bryan was there; I forgot to mention him. He conducted the meetings. Mr. Bryan said nothing, that I know of. That was about the only change that occurred after Captain Hall's death. Everything was kept up as Captain Hall had ordered it. I heard Captain Buddington say that he could not force a man to go to prayer-meeting against his will, and that he would rather take a tramp around—take a walk. Some of the men liked the discontinuance of them, and some of them did not. During the winter season the men used to amuse themselves by taking sleigh-rides. Captain Buddington gave us all the privilege of going off on our own hook at times, and allowed us the use of his harness and his dogs to ride out with as often as we pleased. We frequently went north nine or ten miles from the ship along on the coast close to the ice. No officer accompanied us, and no Esquimaux. We could drive the dogs ourselves. We saw plenty of open water during the winter. The whole of the straits were open there; the ship was frozen in, however, so that it could not get out into open water. There was a bright, clear moonlight; it was sufficiently light for us to see. From Cape Lupton to the ship was clear, open water. There was a little bit of a bight in the land that was ice; we drove along that. Then one day the doctor, Robert Kruger, and myself were out there, and Herman Siemons was attending to the tidal observations. The doctor told us that the flood stream came from the northward. At half past 1 the tide altered, and the ebb-tide then came from the northward. Afterwards I heard that he contradicted that. I saw it every day. I heard some of them say it was not so, but I know that it is. I was observing the tides; I observed the tides for three months. There was a regular ebb and flow six hours each way, with six feet rise and fall. The ice would flood to the northward with one tide when there was no wind. The tide would take a regular change one way or another. When there was a gale coming, by and by the icebergs in the middle of the straits would not take such a fast move to the northward. That is the reason some of them said there was a general current from the northward down south, but it was not so. I am sure the current was the other way, from south to north, and quite regular. In the spring of the year the melting snow from the ravines and the land drives the water down, and it may run more one way—more to the southward than it does to the northward. In the spring of the year the wind is always from the northward, but not in winter when there is regular weather. The ice out in the middle of the channel would go to the northward just as well as to the southward. It would have its regular hours for the turn. The current was that of the tide. The ebb stream is not strong as the flood-stream. I never watched if there was a constant set in the middle of the channel, that is, while in Polaris Bay. I heard the doctor say there was an ebb-current in Polaris Bay, a little ways outside of Providence Iceberg. Providence Iceberg lies about two hundred and fifty yards from the beach; about one hundred yards outside of that we could see the regular tide north and south as far as we could see anything. I always took the berg farthest out into the stream and took a landmark on the west coast over to Lady Franklin Bay, and watched closely whether it was so or not, and by that land-mark I observed this ice in the middle of the channel set both north and south when it was regular weather and the wind did not interfere.

In the spring of the year we went inland sometimes for eight or ten miles, and went hunting, but we could not see any living thing except the tracks of the leming. We could see these everywhere, but not the lemings themselves. There were any number of tracks there. In the latter part of March I shot the first rabbit. Then the same day we shot some partridges, about nine or ten, I believe. When we had nothing else to do we amused ourselves just as we liked. Finally Mr. Chester ordered the two boats to be made ready to go north.

We had two small whale-boats on the shore, and we took them alongside of the ship, and the carpenter and Mr. Chester commenced to fit those up, put lockers around them, &c., and had them made serviceable. They were intended to travel the straits with, as soon as they were opened; but really the straits were not frozen up at that time—not in March. Mr. Chester's intention was to go with the boats to the northward. Captain Hall had said, when he came back from his sledge-journey, in the fall, from Cape Brevoort, that there was not much use in sleighing on the Greenland coast, as the land lay too much to the eastern side. He said that we would have to go across the straits, and go over to the other coast. When the boats were completed and everything in readiness Mr. Chester and his party and Captain Tyson and his party started. I accompanied Captain Tyson. We were gone about a month. Captain Tyson said that we went thirty miles. Mr. Chester afterward determined that it was only twenty-five miles. Captain Tyson, I think, was not able to take observations; he couldn't correct his instruments. We came into Newman's Bay eight days before Mr. Chester. He lost his first boat, and it took him eight days to construct another and get to where we were. In the evening, a week after that, when Mr. Chester, Mr. Meyer, and he came, at 7½ o'clock, Mr. Meyer asked Captain Tyson what latitude we were in, and he said 82° 02′. The next morning Mr. Meyer and Mr. Chester took observations, and said that Captain Tyson had made a mistake. He said, "No, O no; you had better look at your books," &c. They made it out finally about 81° 55′ 45″.

Then Captain Tyson said that as Mr. Bryan had made his instrument all right, it was in a condition to take observations right away; there was no indexing to be brought on at all, or anything else like that. We did not go any farther. If we had had another captain in our boat, we could have gone across to Cape Union. The whole of the straits were open; there was not a single piece of ice in the way. Mr. Chester did not go over, because he had only a canvas scow; he had lost his first boat. As I said, there was not a single piece of ice to prevent us moving farther at this time. Mr. Chester thought that Captain Tyson would get across with his boat to Cape Union sure. He said he would have gone if he had a boat like ours. We did not go any farther, however. We laid there a month waiting, and then Mr. Chester sent two of his men to the ship. We laid within a mile's distance of each other; Mr. Chester was a mile farther to the northward. We did not want to go back to the ship, and so Mr. Chester sent two of his men back to get more provisions, so that we might wait until later in the season for a better chance for traveling. Mr Chester thought that we could do something a little later. Captain Tyson said it was too late at that time to do anything. When the two men got to the ship, Captain Buddington kept them on board. He said that his ship was leaking. He tried to get up to us with the ship, we being only twenty-five miles off, or twenty-two. He tried two or three times, but did not succeed. I did not see the ship, but that is what these men whom Mr. Chester sent told us afterward, when they came with the provisions. We heard the sound of a gun. He had a double charge in the gun. We could not see the ship from where we were, but we heard the sound of the gun in the evening. Then he told us that the ship was in a very bad condition, and he had been using the big pumps to save coal. He pumped with steam before we left with the boats. Then we went back and left our boats. The captain sent Hans with a note, requesting us to come on board. I could not see what was in the note. I do not want to say what I did not know. However, there was a note addressed to Mr. Chester, and none to Captain Tyson. Mr. Chester told Captain Tyson that the captain wanted us to come aboard, and bring the boats along. Then Captain Tyson said we had better go, and we pulled our boat over the ice four or five miles. Mr. Chester's boat-crew gave us a hand; notwithstanding, it took us seven hours to get the boat on shore. Then we went back and got our clothes, stockings, &c. It took us two days. We then started for the ship. Mr. Chester stopped and laid right in the mouth of Newman's Bay. He had orders, he said, to bring the two boats with him, and he had no chance to come down; there was too much ice.

We could not take the boats back to the ship. We could not have gone over to Cape Union at the time we left to come back. There was too much ice. The ice had come down in the mean time.

We got back to the ship. She lay in Polaris Bay at anchor. We went on board and found the ship was in a very leaky condition. Then they kept up steam and pumped all the while with steam. We heard that the captain wanted us to return from Newman's Bay, in order to save coal by having us help at the pumps. We were ten or twelve days on board before Mr. Chester came on board with his men. He sent some of his men before him—Mr. Meyer and two others. One came the very time that Captain Tyson's boat-crew came. When Mr. Chester came on board the next morning we started the deck-pumps and stopped the steam. Then we were waiting to get a chance to go up north, at least as far as these boats were, so that we might get them. Captain Buddington said that he would endeavor to get farther if he could, but the effort would be to get as far north as these boats were at all events. The ice opened a little and by hard squeezing we got the ships squeezed on to the beach and remained there. For a day or two we were not afloat at all. On the 12th of August the ice opened and we steamed down south. We did not go up north to the boats. We could not. I heard Mr. Chester and Captain Buddington talking about going north, and he said they would try to get as far north as possible if an opportunity offered, but then there was no chance, and we thought if there was no way to get north we had better steam home. It got to be so late—it being the 1st of August. Another man 'and myself, Robert Kruger, asked permission to go and get some clothes that we had left in our boats. We got permission, and we went and we stood about twelve or fifteen hundred feet on a high hill and on the southern side of Newman's Bay. We there saw the farthest land that has ever been seen.

Nobody has ever seen that land but Robert Kruger and myself. It was behind Cape Union. It ran northeast by east. Standing at the southern end of Newman's Bay, and looking due north, just a little to the right of Cape Union, we saw the land running northeast by east to the northward as far as we could see. We lost that land from its running behind Cape Brevoort. We were about a mile on the south side of Newman's Bay. We looked up due north, and saw the land of Cape Union, to the right of it, until it was lost to us behind Cape Brevoort. It seemed to be the same height that the land is abreast of Newman's Bay. We could see snow in the ravines coming down the steep coast in this new land. We saw it when we came down for about ten minutes or quarter of an hour. We had not seen it before; we had seen appearances of land there. Captain Tyson, and the others, also, called it Fly Away Land. We thought before that it was land. Some of them said it was a black sky extending over the open water, and said it was an open Polar sea; but we saw it, this last time, just as certainly as I have ever seen anything. I saw nothing of land, to a certainty, when I was up at Crow's Nest, while they were having a consultation below. I did not report any land at that time, though I thought I saw some on the east side.

When we were up at Hall's farthest point, I saw the north cape of Hall's Land. I was up in the Crow's Nest and I saw, to the northeast of Hall's Land, the other land — high land away up in the northeast, as far as I could see. I did not see any other land, only the big bight that went in from the cape that is just above Repulse Harbor. It was not a very clear day when the ship reached its highest point, though I could see a great distance. I could see water for twenty and twenty-five miles at least, and could see across the straits. This was on the shore of Newman's Bay. It was a beautiful, bright day when I went up for my clothes, where we had left our boats, and when we saw this farthest point of land. Mr. Meyer went out to see if he could see it, but he came back and reported that he was not able to see any such land as was described by us. He said he was not willing to mark anything down that he had not seen himself. When he went out, however, it was a very foggy day.

Without concluding the examination of this witness, the Commission adjourned until Monday morning, at eleven o'clock.

WASHINGTON, *October* 20, 1873.

Examination of HENRY HOBBY resumed:

After seeing the land, which I described on Saturday as being the most northern land seen by any one, at the time when I went back to Newman's Bay to get my clothes, I returned to the ship. The ship laid about ten days or more in Polaris Bay. Then we steamed down south. We stopped in Polaris Bay that length of time because there was no opening by which we could get out. On the 12th of August, at dinnertime, there commenced to be open water and slack ice in the straits. We got steam up and steamed down south. We were getting along well as far as 80° 2' north. There the open water stopped, and we ran the ship till she got beset in the pack. We drove slowly down along the western shore, the ship all the time setting slowly more over to the Greenland side. A house was built on the ice so that we might have shelter in the event of our losing the ship. We had some provision in the house. Captain Tyson had that done. We drifted down fast to this ice, until the night we broke adrift in the gale. We could not have gotten on to the western shore at any time after we got fast in the channel. We were blocked in. After we passed Dr. Kane's winter-quarters we drove down rapidly. Passing Cairne Point we saw plenty of open water. The trouble, however, was that the ship was frozen in to one floe. The open water was to the northward, to the southward, and east and west of us. We, however, were frozen in to one big floe, and there were some other floes in the pack where we were. We drove past Cape Alexander and saw Northumberland Island from the deck. A heavy gale commenced from the southwest. It was snowing and blowing fearfully for about thirty-six hours. In the evening at 6 o'clock it was dark, and all at once a heavy crack in the floe came. Two of us jumped up on deck. We saw the floe was parted right where the ship was lying. In about five minutes all the ice was gone on the starboard side. Captain Buddington called all the hands to get the musk-ox skins, provisions, &c., ready. We did not commence right off to heave overboard. We had provisions on deck and coal. The musk-ox skins and clothing we had to keep dry in the room. We got them up in port-alley way. We waited a little while, and all at once we got a nip on the starboard quarter. A big floe or a little berg struck her there fearfully, and keeled her over on her port side. Then Captain Buddington sung out, "Heave up all you can, as fast as possible." Some of them went on the floe, and others threw the stock overboard, forward and aft. Three of the party brought the things forward to us, and Mr. Chester and I hove them overboard. Captain Buddington, Mr. Morton, and some others were aft. We had the pemmican aft on the poop deck. He threw that overboard. I was forward along with Mr. Chester. We did not heave everything overboard that was there. Then the men sung out that they wanted the boats lowered, and we lowered two boats down. They did not like to stay, they said, any longer without the boats. When we had the two boats down, we commenced heaving more provisions on the ice. Then Captain Buddington and Mr. Chester spoke together, and said it was better to stop a little while, and not to heave everything over just then. We waited for a couple of minutes, and then Captain Buddington said that I should go over on the floe, and should carry the provisions that were on the floe-edge, alongside the ship, out on to a higher part of the floe, where we had the house. I just went over the rail, and went down the lower part of the steps. Through the heavy nips the ship had had, the floe was broken up, and there was about three fathoms there broken up in little lumps of ice. I was, therefore, unable to get off. I sung out to the captain that I could not get off that way. He said you had better come up. At the same time the steward came running up. He was on the floe. I stood on the lower part of the steps; and he says the floe is broken all over, and I must come on board. The captain says I have just ordered a man off, and if he cannot get off, why you cannot get on board. He said I must; the floe is broken all over. Just at that minute off went the ship. The fastenings broke, and the ship went off. There we could see three-quarters of the provisions on this little piece of ice yet, and there were about four or five men on it. Some of them were on the better part of the floe. Those that were on this little floe were shouting, and saying they wanted to get on board of the vessel. The captain said, " I have got no boats on board, and I wish I was where you are." We all wanted to be on the floe. We had no boats, or anything of the kind, and the ship was in a fearful condition. We drove away, and that was the last we saw of them. The steward sung out, "Good-bye, Polaris." Those were the last words we heard. That was about half past 9 in the evening. At 11 o'clock our ship laid still, I think. We could not see any land. The snow was blowing, yet it had calmed down a little. All around the ship there were broken pieces of ice. The next morning at 9 o'clock, we saw the land on the Greenland side. We saw that we were about three miles from the coast. At 9 or 10 o'clock, I went up to the mast-head, and I saw a lot of provisions about four or five miles from the ship. I could distinguish the coal-bags, the boxes, &c., one from another, but could not see any boats, any house, or any living man, and no dogs. I was looking through a glass. I saw these things from the deck also, and when I came down I could show it to the others around. The floe was a little to the southward; more to the southward than abreast in the straits from us, a little to the southwest of us.

Mr. Chester went up and saw the same thing that I did. Afterward Captain Buddington sent me up again, but I could not see any men. I went right up as high as I could get on the topmast with a glass, and I could not see any movements or anything of that kind. This was about 12 o'clock in the day. Then we had gotten steam up in the little boiler. We had steam up, but the captain ordered more steam in order to use the propeller. We had got steam up to pump the ship with. Then there was a little bit of a lead opened into the shore. A slight breeze sprung up from the northeast. A little bit of lead opened into the shore and we tried to get in on to the beach. It took us until 4 o'clock in the evening to get on to the beach, working with sail and steam, but there was no other open water to be seen anywhere except a little to be seen toward Littleton's Island. We made toward the main land, between Littleton's Island and Life-Boat Cove.

She was beached where the fall of the current swept along. I would have beached her two hundred yards farther down south, around the point.

The next morning we commenced to take the spars, topmasts, yards, &c., down, and take them on shore. We took all the provisions we had left on shore. Two days afterward there came some Esquimaux to us. The captain gave them several things, and in the evening they went home. The next day there came six of them. They helped us to take all the stuff on shore. I cannot think of anything just now that happened during the winter. We remained there, however.

In the spring the doctor wanted me to go to the North Pole with him on a sledge journey. I thought it was a very foolish idea, with fifty pounds of pork, and sixty pounds of bread on one sled, to go to the North Pole from there. At this time we were two hundred miles farther south than we were the year before, and yet we did not try it then, when we were farther up. I was told to go, however, and I said I would go; that it made no difference to me. The doctor promised me $100 to go to Thank God Harbor, and $200 if I would go with him so that he could reach a higher latitude than Captain Parry reached. His principal object seemed to be to go to Thank God Harbor. What he was going to do there I could not say. The captain bought a team of dogs and sleds from these wild Esquimaux who came there, and we had those at the time. Dr. Bessels was constantly speaking to me about going to Thank God Harbor with him, and we had arranged for such journey, but before we were able to start, the ice broke and the journey was accordingly abandoned.

We lived in the house that we built all the winter. A fortnight before we left, a gale of the northeast took her about one hundred yards farther south, and then she broke out in open water. The hawsers were parted, and Herman Seimens and I made one of the hawsers fast to her again. She only was twenty yards from the beach. Before, she was about one hundred and fifty yards from the beach. She was three-fourths full of water at the time. The high tide forced her up on to the beach. We made fast only a single hawser. We were not told to do this, and we had perhaps no business to do it, but we took the responsibility. If I had had anything to say I would have secured her properly at that time. When we went away with the two boats that Mr. Chester built, we passed right in front of her stem, and she was lying there level with the water. We sailed down that day in sight of Littleton Island. We passed Cape Alexander and down to Etah, the second settlement from the north. Then we traveled along very comfortably. Nearly all the while we had abundance of water. We got pretty rapidly down south, and on the 23d we saw the whaler Ravenscraig. Some of the party were glad to see her, while some were very sorry. We did not want to go across to the other side for two or three months. We thought we could get to Tessiusak in ten or fifteen days. We had made more than half of our passage down in twenty days, and had six weeks' provisions more in our boats, and everybody was in excellent health. We thought we could get down there sooner, and, as we had plenty of provisions on board, would have preferred remaining on the boats. I wanted to go to Disco, and all in our boat wanted to go there. I am sure we could have reached Disco without any difficulty. Mr Chester did not want to go on board the whaler, but Captain Buddington did. We had not eaten more than one-third of our provisions at that time.

At the time our ship went adrift from the ice-floe and we were separated from our companions, she was in a fearful leaky condition. I was down in the fire-room, and saw that the water was just coming on to the fires.

We had to start the deck-pumps as quickly as possible in order to prevent its doing so. We had to thaw them out first with hot water from the boilers. We must pump with the deck-pumps or the engineer said the fires would go out. Everybody was working as hard as he possibly could. The stem of the ship was gone; the six feet of it was broken out. I could stand right in the hole. When the ship went on Providence Iceberg on the 22d of November, 1871, Captain Buddington said that it was the safest place we could have her. All of us said the contrary; I had never seen a ship setting on the ground the whole winter, and this was the same; she soon commenced to keel over a little, and kept keeling over a little more and more all the while. She was leaking fearfully all the winter. When she had set about a fortnight, then the captain thought it would not be a good plan to leave her there during the whole winter. It was then, however, too late. If he had come to this conclusion before, she could have been gotten off in about an hour's time. We could have sawed her off. There were only two or three inches of ice on the port side. That is where she got her break in the stem. During the winter she was cracking sometimes fearfully. In June, of 1872, we found out that the ship was leaking badly. I knew well enough before that it could not do anything else; that it must be leaking. That discovery was made when we commenced thawing out of the ice. She had broken the stem; the piece did not come out at that time. We tried to fix it. The big planks were broken right in the middle square off, and the stem was bent regularly away for two or three inches. After we ran her on shore the morning after we parted from our comrades, as we were walking over the ice taking things on shore, I went forward to her stem and saw how she looked there. The lower part was all away at the six-foot mark. A large portion of the stem had come out the night that we parted from the floe. She was built up solid forward, otherwise she could not have floated after that. She was a strong ship and a very comfortable one. She would have been right enough if she had not been kept on that iceberg all winter. The nip that she received the last night would not have done her any harm. She would have stood that all very well. This break on top of the old one caused her destruction.

Q. How often did you go to the mast-head to look after your companions?

Answer. Twice. I staid there from ten minutes to a quarter of an hour at a time. Everybody was very busy taking care of the ship during that time. The first time I was up there I was called by the captain to attend to the other business. It was while they were eating breakfast that I went up to the mast-head. The first time I went up on my own hook; afterward I was sent up by the captain. The time that Mr. Chester and I were up at the mast-head altogether would make about an hour and a half in that day. There was nobody looking from the mast-head about 4 o'clock in the afternoon. I think the reason why we did not see them was that they must have been behind Littleton Island, or in the shadow of some berg. There was no men on board the Polaris that could beat me in seeing. That is what they all admitted, that I had the best eye-sight. I was regarded as the best look-out on board. I could distinguish a bird while it was flying, and see whether it was a bird or something else, when the others could not tell what it was. I have very fine eye-sight. I do not see particularly well through a glass; I can see with my bare eyes better.

There was a great deal of refraction in the atmosphere in that region. I saw a great many mirages, and have seen ships and land lifted up which would otherwise be below the horizon, and not visible. I do not know that there was anything of that kind the day that the other party saw the ship. I think that if they had been anywhere near the floe that had the provisions on, I should have certainly seen them.

The land that Mr. Kruger and I saw farthest north, when we went to get our clothes where we had left the boats, consisted of very high peaks, coming steep down to the coast, apparently. We could see the snow-ravines between the peaks. There did not seem to be any capes there, but the land seemed to come straight down as we saw it. It lay off northeast by east, and not up toward the west coast from where we saw it at Sumner Headland, at about an elevation of twelve hundred feet. We could see it until it stretched off to the northeast, and was lost behind Cape Brevoort.

It was a splendid day when we saw this land. When we were on the boat excursion we had been in there for a month, and we were. not always able to see Cape Union, but at this time Cape Union looked as if we could heave a stone on it, it was so near. I had no glass. Kruger had Mr. Meyer's spy-glass.

When we first went up in the Robeson Channel the ship could have gone farther north. After they had the consultation on the house, which was on the 29th of August, I think, we could have gone farther north. And I think we could have gone farther north after we got back into winter-quarters. We had a gale from the northeast, and we drove down with that from our high latitude to 81° 38′. That cleared out the channel, and we could have gone on north after that. I am quite sure we could have gotten as far as we did before, at least. There was then blowing a little breeze from the northward, but not much, and there was no ice coming down at that time. That, of course, showed that there was open water. The reason we did not go farther north was because Captain Buddington said that it was not safe to go farther north; that we had not left any depots of provisions anywhere. He said we were a thousand miles from our first depot. Mr. Chester wanted to go to Newman's Bay, at least. He was waiting the whole day to heave anchor to go north. That was the first day we lay at Polaris Bay. Captain Buddington said this that I have just repeated, about not going any farther north, forward on the deck. It was spoken in the presence of every one. When this consultation on the house was had, Captain Hall, Captain Buddington, Captain Tyson, Mr. Chester, Dr. Bessels, Mr. Meyer, were present. Captain Hall asked every one of them, and they all said they wanted to go north, with the exception of Captain Buddington. Captain Buddington wanted to go into Newman's Bay and make winter-quarters there. The others wanted to go north. Tyson wanted to go into Newman's Bay also, if they could not get north on the west side. The rest wanted to go north. There was open water to the northward at that time. The ship was then lying in open water. Some of them wanted to cross the straits, and some of them wanted to keep the Greenland coast. The open water was on the Greenland side. Some of them said it would make too much easting, and the others said that she never would get across those straits; that she would get fast into the pack. That is what Captain Buddington said. We tried to get across the straits on the other side, and did get fast.

I remember when Captain Hall was sick, and when he died. I was not with him at all; we were not allowed to be with him. I saw him only once, and that the Sunday when he was quite well. He died in two or three days after that. I only saw him the one Sunday I have spoken of. After his death I never heard any one express himself as

being relieved by his decease. I know there were a couple of officers
who were greatly relieved by his death. The doctor was one of them
that I know of. I think Captain Buddington was also. I never heard
them say so: I could see it by their works. One of the officers said
that now they would have something to say; that before the sailors
had the command; that Captain Hall consulted with the sailors, and
not with his officers. They said that they would find it a little different now. Mr. Meyer said that. He remarked that now the officers
would have something to say; that they had nothing to say before. I
know that the doctor was greatly relieved. He did not know what to do
when Captain Hall was alive. When Captain Hall would call one of
the scientific men all three of them would jump up, and each one would
suppose he was called on. Some of them did not want to behave very
well. Captain Hall said he would court-martial the doctor if he kept
on in the way he was doing. Nobody ever said in my presence "that
there is a stone taken off my heart now," referring to Captain Hall's
death. Captain Buddington said to me at one time, "We are all right
now." He said that the very same morning that Captain Hall died. I
said, "How do you mean about that?" He said, "You will have plenty
to eat now, and you shall not starve to death."

The discipline of the ship was good during Captain Hall's life-time—
first class; afterward it was not very good. The most I can say of
Captain Buddington is, that he knows how to manage a ship in a first-
class way, and is a first class ice-navigator; also Mr. Chester.

Examination of Hermann Siemin.

I am a native of Germany, and thirty-one years of age; by profession
a seaman.

I have sailed from Germany, England, and from America. I passed
the examination before a board as a ship-master in Germany, in 1868.
I never commanded a ship. My highest rating has been first officer on
board of a Nova Scotia sailing-vessel that sailed out of New York,
named Eolus, Captain Perkins. I never sailed in the Arctic circle
before the time of the Polaris expedition. I joined the Polaris in Washington. Nothing remarkable happened until we reached Tessiusak.
Tessiusak was the most northern port we made. From that point we
went directly north. The first place at which we landed was with Captain Hall, and Chester, and four men, of whom I was one, at Cape
Frazier. Captain Hall could not find any winter-quarters there. Then
we went farther to the northward, and by Cape Lieber one day he
stopped the engine, to get proper observations to ascertain in what
latitude Cape Lieber was. He had the whole scientific corps and the
officers on board to take observations in order to discover this exactly.
Then we had to stop the ship for a time at the place, because the weather was so thick and foggy that we could not see her course. One
morning, I do not remember exactly the date, we got our highest latitude at about 6 o'clock. I went myself in the crow's-nest to look out.
I was asked if I saw anything to the northward. Farther to the north
I saw no lead, nor did I see any prospect of getting any farther north.
Captain Hall then concluded to look for winter-quarters. In about 82° 9′
north latitude we looked for winter-quarters, but the current was a kind
of a maelstrom—it turned around and around, so that there was no
place there for the ship to make winter-quarters in. This was above

Newman's Bay, at Repulse Harbor. Then we went farther down, but could not reach the coast, and could not find any better place. We were looking toward the west coast for winter-quarters, because that land was more to the north than the east coast was. Greenland turns too far to the east, but we could not reach the west coast. We drove down from the latitude that Captain Hall said was 82° 26′ until we reached winter-quarters in 81° 38′ north.

The next day we were engaged in looking toward the southward for winter-quarters, but we had to return. Thank God Harbor is not a bay, but only a bight, and there is no shelter there from the southerly and southwesterly winds. But we could not find any better place, and so we returned in the evening to our old harbor again. We then commenced putting the provisions and stores on shore as quickly as possible. After awhile, Captain Hall made a sledge journey to the north. When he came on board, so they tell me, he stated that he had not been in very good health during the last three days of his journey. I saw him when he came on board, but did not speak to him. I did not speak to him at the time, because I was one of the tide-observers, and Henry Hobby and myself were working the snow out of the tide-pond. Some of the officers said that he was not well the last three days of his journey. I don't know who it was that told me he hadn't been feeling well, but some one told me so. I did not see him during his sickness. I asked Captain Buddington for permission to see him, but never had the privilege. He told me that he would see what he could do for me. That was all the information I got from him. I did not see him at all while he was sick. I asked Dr. Bessels about Captain Hall, and he told me that he would not get over his sickness. This was after he had been taken sick, but before he got so very sick the second time. After he died we buried him. That was all, really, that I know about it. I never heard any formal announcement of who had command after Captain Hall's death. Dr. Bessels told me that everything would be the same as it had been, with Captain Buddington as sailing-master; still, we lost everything when Captain Hall died. I mean by that, that in my opinion the expedition died with Captain Hall. No ship ever had the privilege we had. I was with Mr. Chester twice in his boats. He did what he could, but the thing I did not like was, there not being any sledge journeys. The straits in the spring were so that we could have crossed them toward the west coast of Grinnell Land. The ice was not moving at all. But it was not done. I have no doubt if Captain Hall had not died that we would have reached the highest land on the west coast that Mr. Meyer has laid down on his chart. We could not have gotten any farther north with the ship than we did go, I think. As I stated, I accompanied Mr. Chester in his boat expedition. At that time, we got in the mouth of Newman's Bay. I did not go any farther except that I went on the shore with a telescope, and went some miles behind Cape Brevoort. I looked to the northward, and I think I saw land, but I cannot say for sure; but my belief is there is land there. This land that I speak of lies across above the land laid down on the chart, and stretches from west toward the east. With the telescope, I could see an opening from between the coast-line on the west side, as laid down on Mr. Meyer's chart, and the land which I saw stretching off and running off toward the east. The land was so far off that no one could see exactly how it lay within a point. This land lies behind a steep cape not laid down on the chart. From the highest point on the east coast, as laid down on Mr. Meyer's chart, I saw still to the northeast of that a high cape with

water between that and the point put down by Mr. Meyer. I could not
see whether it was a bay that lay between the two capes, or whether it
was a sound. I could not see land to the southeast and do not know
whether it joined it or not. This farthest land which I saw was still
farther north than this cape I have spoken of, and ran off behind it
with, I think, water, and an outlet between, toward the northeast.
The glass I had was one of Mr. Meyer's large telescopes; a heavy
one; one man had as much as he wanted to do to carry it over the mountains. I could not see this land with the naked eye. It was a very
fair, bright day when I was there making this observation. I told
Mr. Meyer of it when I came back, but he never saw it himself.
When I came down from the boat journey they were looking for a
chance to go toward home. Mr. Chester staid on his boat journey as
long as he could for the purpose of seeing if he could not find an opening. I believe if he had seen an opening that he would have gone to
the north with the boat we had. There was, however, no opening.
There was not even sufficient to take the boat down to the ship, so that
we had to leave the boats and everything there. Then, on the 12th of
August, we left Polaris Bay for home. We got a little lead, and pushed
our way through the ice. Thus we steamed so until we reached eighty
degrees and two minutes north, where we got beset in the ice. Then we
had to make our ship fast for the first time in coming down; we drove
in about a month and a half. I do not remember exactly the time
when we got beset, but we drifted down in the pack-ice until we got
this smashing up, when the Polaris got stove in; we drifted down the
pack-ice until the 15th of October, when we got a gale from the southward, and in the evening about 6 o'clock the ice separated from the
starboard side of the vessel. About half past 9 o'clock, after the ice
separated, it came in again on us, and nipped us. When the ice came
in again, Captain Buddington gave orders, while the ship was cracking
all over, to land the provisions on the floe. Some of the men went on the
floe to transport the provisions; at half past 9 we broke loose and drove
away, and left the party on the ice. We drove for some distance in open
water,and then brought up in slush ice that would not bear a man's weight.
It had been made within a day or two before in the bay. When we drove
away from the party the water was beginning to come rapidly into the
ship. We had to take the hot-water out of the small boiler, so as to make
the big deck-pumps work. We had to thaw the ice out of them. The
engineer in his department was below, and made the fire up. He burned
blubber, wood, and everything that he could get, to get steam up to run
the pumps, so that we would be able to pump her with steam. Just before the pumps were working, Mr. Shuman told me that the water was
nearly up to the furnaces, and came very near putting the fires out.
After he got the pumps to working by steam, we got the better of the
water. Thus we were enabled to keep up until the next day. We then
worked our way toward the coast of Greenland with lines, sails, and
sometimes a few turns of the propeller, until we reached in the evening
the coast. But we could not get upon the ground for the ice which lay
on the shore. We got as far as we could, and when the tide went out
the vessel was on the ground. The next day we had worked the ship
still closer into the shore, as close as we could get her, and then fastened
the hawsers to the hummocks that were aground, so that she would not
drift off again. Then we commenced to bring the provisions on to the
shore, and we took the spars down and brought them on shore. The
next day Mr. Chester turned to with a couple of men and commenced to
build a house. The Esquimaux came with five sledges and assisted us

in getting the provisions and everything we could from the ship on to the shore. After we had everything out of the lower hold, Captain Buddington gave orders to let the engines stand. There was then no more pumping with the engine, and we were therefore forced to let the Polaris get full of water, because the stem was broken off at the six-foot mark, and totally away from the ship, so that a man could stand in the opening where the piece of stem had been. Even the boards and sheets of iron were bent out. We fixed our house on the shore, and tried to make ourselves as comfortable as possible, and there we lived through the winter.

The next day after we parted from our comrades we kept a sharp look-out for them. The chief officer was in the "crow's nest" the whole of the forenoon. We were not able to see any of the men. Mr. Chester went into the "crow's nest" with a glass, and looked around and around for them, but could not see anything. As far as I know, there was some one at the mast-head in the afternoon. I cannot say whether anybody was at the mast-head at 4 o'clock in the afternoon or not; but Mr. Chester, while we were going into the shore, was up in the "crow's nest" looking for leads, at the same time that he was looking for the men.

I kept a diary. My first diary was published in the other report. I kept one afterward. These diaries were written each day as everything happened, and will give my story more particularly than I can remember it now. Of course, a man cannot remember everything that occurred two years since, and I would not like to say before God and my Government what is not exactly true. We started from our second winter-quarters at Life-Boat Cove the 3d of June. The boats had been made under the directions of Mr. Chester. We made our way down about twenty-five miles below Cape York. We had pulled upon the ice, and were watching for a lead when we saw the Ravenscraig fast to the ice-floe of Melville Bay. We were taken on board of her, and went with her across the channel, Lancaster Sound, over on the west coast. We afterward went on board the Arctic, and, when she was ready to sail, we went with her to Dundee. Three of our comrades, Mr. Bryan, Mr. Joseph Mauch, and Mr. Booth, were left on board of another whaler called the Intrepid. The whalers leave the whaling-ground about the middle of October, and it is nearly or quite time that we should hear from them. If we had not fallen in with the Ravenscraig, I think we would have gotten down to Upernavik or Disco in our boats. If we had not succeeded in reaching there, we would probably have been picked up by the Juniata or the Tigress.

In our first winter at the north I found on the south side of Newman's Bay, a mile and a half inside of Cape Sumner, in a distance of a quarter of a mile, twenty-four pieces of drift-wood.

With a northerly wind they would have just come from that opening that I had seen to the northward, and which I have described as being between the west coast and the fartherest land which I saw.

They would sweep just clear of Cape Union and drift on to the southern coast of Newman's Bay. We burned some of this wood to boil our coffee with and cook something to eat. I cannot remember the size of these pieces, but that is given in my diary. I would remark that in the statement of the length as given in my diary it is not stated whether it is feet or inches; but I meant to have it inches. Some of the men took a few pieces of this wood on board the ship and gave it to Dr. Bessel. The rest we used up or left. I cannot say what wood it was. I believe the doctor had a name for it. It looked to me like hard wood. (Pieces of wood exhibited by Dr. Bessels while giving his testimony were here shown Mr. Simmons, and he said, "These are the same kind of wood, and look like some of the pieces.") The greatest length of any of the pieces I found was about eighteen inches. At Polaris Bay we found musk oxen, rabbits, lemings, some birds, in the summer time, of different kinds, and got one white fox.

As regards vegetation, there was a kind of grass there. I do not know what the name of it is in English, but I should call it in German "heide." We came across little grass plains, and met with flowers in the summer time. During the summer season the land was pretty clear of snow, with the exception of some deep ravines.

I saw the track of the glaciers in Newman's Bay, and I have even heard a glacier discharge. I heard one discharge below our winter-quarters where the place called Southern Fiord is. I have seen stories in the papers about Captain Buddington's drunkenness, but I have never seen him so drunk that he could not discharge his duty. He is a sailor, and a splendid ice-navigator.

I will state that the more particular details of what I know will be found in my diary. Statements therein contained were written down by me every day as the circumstances occurred.

———

Examination of Alvin A. Odell.

I was born in Connecticut; shipped as second engineer on board the Polaris at New London; sailed with the ship from New London on the 3d of July, 1871; went with her to St. John's; thence to Fiscanaes, thence to Holsteinberg, thence to Tessiusak, and thence northward on my northern voyage. After we left Tessiusak we proceeded to the northward. I was in the engine-room most of the time, and was not, therefore, very familiar with what transpired on deck, and of course did not see as much as those who were on deck. I sometimes ran out, and what I saw was at those intervals. We proceeded to the northward through the ice as best we could. The particulars I am not able to state, for the reasons which I have before given. After we had got along for some days—a few days before we went into winter-quarters—we got beset in the ice in a strait which Captain Hall called Robeson's Channel. After we got beset in that ice, we got out afterward, and went up still farther in the channel above a cape, which he called Cape Lupton, and a bay, lying eight or ten miles above it, which he called Newman's Bay. We afterward drifted down in the ice from the highest point we reached, which I understood to be, after the latitude was corrected by a scientific observation, 82° 16′, to the point where we went into winter-quarters, at a latitude that was said to be 81° 38′. After we had gone into winter-quarters we landed our provisions on shore, and set up an observatory. We banked the ship in with snow, and covered the house with canvas, and made ourselves snug for the winter. About the 10th of October Captain Hall made a sledge journey to the north, accompanied by Mr. Chester and the two Esquimaux, Joe and Hans. While he was gone, we were engaged in making ourselves comfortable in our winter-quarters. He returned about the 24th of October. I saw him when he came back at the gangway. I shook hands with him, and he wanted to know how we did. I told him "pretty well." I told him we were banking up the ship. He said he was glad to hear it, and smiled, and went in. He said he was pretty tolerably well, as far as I

9 P

could understand. There was not much said, but from his looks I
thought he was quite well. After that it was but a little while before I
heard he was sick. What was done I do not know exactly, but I heard,
after a while, that he was getting worse, and that he kept getting worse.
Some little of the time, I believe, he was better. In a few days he died.
Shortly before he died I looked up in the scuttle, and I saw that he was
walking up and down, and I thought from that that he was getting
along nicely; but the first thing that I heard was that he was dead. I
was not in the cabin with him during his sickness more than once; he
was sitting up in his chair then. I had a minute's conversation with
him. I do not remember particularly what he said. I did not go into
the cabin again, and therefore did not see him until he died. Captain
Tyson, Mr. Morton, and I laid him out after he was dead. We buried
him on the 11th. I did not take any particular notice, and did not hear
much about it. I have no reason to suppose that he died anything else
than a natural death. Captain Buddington went into command after
his death. After that time we made ourselves as comfortable as possible. We did not do much of anything. Once in a while some of the
party made a sled journey. The next spring an expedition was made in
boats. The scientific operations went on during the winter, as far as I
understood.

Shortly after Captain Hall's death, in a gale, the ship broke out from
her anchorage and we drifted against an iceberg. She was made fast
there, and rested on the spur of this iceberg all winter, rising and falling
with the tide. She would right up a little with the high tide, and as
that fell she would fall over, resting with her stem on the spur of this
iceberg. She strained herself a great deal during the winter, and in the
spring, when we got clear of the ice, we found her in a very leaky condition. We found the water coming into her very rapidly. We got
pumps to working then. We managed to do this by hands changing
off. One gang would take hold and work at one time, and then another.
We did not at first pump by steam. After a while we found she was
making pretty free, and we used the steam-pump.

We had to wait until the men came back from the boat expedition, and
then we worked the pumps by hand, and discontinued the working by
steam. We had been using the small boiler, but worked the big pumps
by hand, thus saving our fuel. After the boats were gone we tried three
times to get to where they were, but were not able to get past Cape
Lupton, though we got abreast of it each time. We had to go back
each time into our old quarters, and finally staid there until we started
to come home on the 12th day of August. We made our way south
slowly through the ice, there being slack ice all around us, and finally
got stopped in the pack again in the middle of the channel. After we
got beset first in going south we got free again, and got out into the
channel, and then got into another pack, and there we got fast. We
tied up to an old floe which was very large and solid, about two or three
miles long. We drifted to the southward and past Cape Alexander into
Baffin's Bay, and were thus situated until about the middle of October.
Then there came that heavy blow and gale and snow-drift, and we were
separated from the floe. The ship got a pretty severe nip, and that
caused her to break loose from the ice. We were thus separated from
our companions on the ice while we were in the act of taking off the
provisions and materials which were on deck, ready to be put on the
floe in case of emergency. I was below when the thing happened. I
was assisting about, and once in a while I would run and help heave
over some things until we found that the water was increasing very

rapidly upon us; then I went to the hand-pumps and commenced assisting there until we got steam into the little boiler again. The water came in so rapidly that it was all we could do with all hands working, to keep the water from the fires. After we separated we had to keep pumping very hard until we got steam up, and then we began to pump by steam. The next morning we saw where we were; at least we made out after a while that we were up by Littleton's and McGary's Islands, within a couple of miles of the shore. The ice was a little slack. We had steam then in the little boiler, but we had to get up a pretty good pressure, and then set the engine to work. We would run the steam all off, and then we would have to stop and get up steam again. We would get fifteen or twenty pounds; then we would put it on; and by that means we ran the vessel on to the rocks; ran her aground, and got all the things out and such provisions and coal as we had on board. The loose ice by the side of us froze together after a few days. We then took the provisions ashore, and had a house built there from the wood we got out of the vessel, bulk-heads, and such other parts. We covered the house with canvas and spent the winter at that place. The next morning after we got adrift from the ice, Mr. Chester went up into the " crow's nest " to see if he could see anything of our other party, but he could see nothing of them. He had an idea that he saw them at one time. He thought he saw something that looked like bags of coal, but afterward he concluded that he was mistaken—that it was nothing but black ice. We wintered there as well as we could. The Esquimaux came to see us, and were friendly disposed, and helped us all they could. In the spring we made boats. Mr. Chester, the carpenter, and all hands set to work doing so. I did what I could during the time. On the 3d of June we left our winter-quarters and went south ; and when the ice prevented us from going, we hauled up till it opened for us. So we worked our way down below Cape York, when the Ravenscraig hove in sight. After awhile we went on board of her, and crossed over the bay with her to Lancaster Sound, and were with her there while she was whaling during the summer-time, till she left in September, until the whaler Arctic came along, pretty nearly full, ready to go home, and we were transferred to her to sail for Dundee. I do not know the date of our arrival at Dundee. We afterward came on to Liverpool, and thence sailed for the United States in the City of Antwerp, and arrived in New York, at the Brooklyn navy-yard. I kept no journal. My duty was below, in assisting at the engine. Of course I did not see as much of what happened as those on deck. I did not have the same opportunity. I have given a general statement of what occurred.

During Captain Hall's life-time the discipline of the ship was very good. Afterward it was not so good. Captain Buddington would get pretty well " set up " once in a while. I cannot say that he was drunk, but he would go around like——Captain Buddington was a pretty easy sort of a man, and rather familiar with the men, and that made discipline rather loose. I never heard anybody say that they were relieved by Captain Hall's death.

I do not think I saw any chance to get farther north in the ship than we did get at any time. The ship was severely injured when she broke from the ice, the time when we separated from our comrades. She was making a good deal of water at the time. It was as much as we could do to get into shore at Life-Boat Cove. I did not see much difference between the temperature of the second winter and the first. There was a little more snow farther south. I do not know whether Captain Hall kept any journals or records, and, of course, I do not know what became of them. We had a very good crew. Everything went on peaceably. There were no outbreaks of any kind that amounted to anything. I do not think that the Polaris was exactly of the right build for a ship to go north; but she was very strong. The machinery was very compact, but a little unhandy; but we got along with it, however, very well indeed. It was in good condition. I do not know of any disagreement between anybody and Captain Hall. I have heard there was, but I do not know anything about it. Captain Hall was a very kind man. He was quick once in a while, but he was a man very easy to get along with.

Examination of Nathaniel J. Coffin.

WASHINGTON, *October* 21, 1873.

I am a native of Portsmouth, N. H. I learned house-carpentering and joiners' trade in Portsmouth, N. H. I worked at ship-carpentering on the Pacific coast some, and in Portsmouth, N. H., navy-yard, and in Washington navy-yard. I shipped on board the Polaris here, at Washington, and sailed with her from Washington as far as New York, where I was taken sick. I was afterward sent forward by the Congress, and rejoined the Polaris at Disco. I sailed with her from Disco to the north. Nothing of importance happened other than Captain Davenport coming on board the Polaris and reading the object of the voyage, &c. There was some little misunderstanding between Captain Hall and Dr. Bessels, 1 think. Captain Hall stated that he had been insulted by Dr. Bessels. He stated that in the cabin before all of us. It was at the time that he read off the duties of every man. I could not say exactly where that was, but I think it was just before sailing from Disco. Afterward we proceeded on to Uppernavik, and then went to Tessiusak to get some seal-skins and dog-skins at those places, and secured one of the guides—Hans and his family. From Tessiusak I think it was we went on north. The first place we came very near, I think, was near Cape York and the conical rocks. From there we bore over to the western shore and went up by Cape Frazer. There was a boat put off there, I think, in an effort to find winter-quarters, as I understood it. After that we got beset in the ice and drifted farther south, so Captain Hall read off in the cabin. He read off either before or after Sabbath service that we were in latitude 82° 26' north, and thought we had drifted into 83°, but was not certain. That was after we got into Polaris Bay that he read this off. We had drifted and come down very near Polaris Bay, and the ice opened. After we left Cape Frazer we passed up through Kennedy Channel, passed Cape Constitution on the one side, and passed up through what was formerly Kane's Open Sea, now called Polaris Bay or Hall's Basin. We passed Lady Franklin's Bay on the west into a channel with land on both sides up to a latitude which Captain Hall called 82° 26'. There we got beset in the ice, and after some time drifted back to the strait. We drifted sometimes backward and forward with the tides. I heard Captain Hall state on board the Polaris that he thought it was possible that we might have drifted to the north into latitude 83°, but he was not certain. While we were beset, we put out a great deal of our provisions on the ice, for fear that we would have to abandon the ship.

When the pack began to loosen, we loaded the vessel, put the things on board again, and when the ice was open 1 believe there was an attempt made to go farther north. We found this Polaris Bay, and then made one attempt to get farther north, but we failed. I remember Captain Buddington saying to Dr. Bessels that we were about two inches farther north, he thought. That was when we got back again into Polaris Bay. We anchored there, and sent things ashore in a boat preparatory to quartering there for the winter. Shortly afterward the ice made around us, and we banked up the vessel. At the time of Captain Hall's death we were banking up the vessel for winter-quarters. The awning was put on, and the banking was nearly completed when Captain Hall died. After we got into winter-quarters, and before the banking was completed, Captain Hall made a sledge journey toward Newman's Bay by the land. He started out to what I thought was the east, but he did not appear to be positive of it from the way he wanted the observatory set. He said he wanted it in just such a position, and then told me that he thought he would have to alter it, as he did not know exactly the points of the compass at that time; that he would have to test it before I opened some lights on the top for Mr. Bryan's transit-stand. He made the journey, and was gone a week. When he came back I had orders to make some wheels. I made three of them, and then I was ordered to discontinue them at the time of his death. The reason why I was ordered to make the wheels, was because Captain Hall encountered a great deal of bare ground, and he wanted to go over that when he could not use the sleds on account of there being no snow. He was calculating upon another journey right off, as soon as he recovered, before the season set in. I cannot tell the day he started; my log will tell, I think. He was gone something over a week, I believe. I saw him when he came back, and shook hands with him, and he appeared to be perfectly well. I saw him at the cabin door; I called to see him. I messed in the passage-way, and had my room forward. I never saw him after the first time, after his being taken sick, but twice. When he was very sick, I made an excuse to go into the cabin to see him; I had a piece of furniture to fix; I took that in, to see him then, and once after that Mr. Morton asked me to come in and open a keg of tamarinds. When I took that chair in to fix it, I had a little conversation with him; I asked him how he did, and he said that he thought he was getting along better; I had no other conversation with him; Mr. Chester was there with him at the time; there was nobody with him but Mr. Chester. The second time I saw him, was when I went in to open the box of tamarinds; I had not much conversation with him at that time: I only asked him how he did; he stated that he believed he was getting better; he was then in his easy chair, with a counterpane and cover wrapped around him, sitting up; both times he was sitting up; I never saw him again till after his death; I heard from him every day at the table. The steward and the cook both slept in the cabin, in the same place that he did; their births were opposite. I would hear statements about his health every day; I would ask if it would do him any harm if I were to call in; they said that they thought it was advisable not to disturb him. I asked Hans what he thought of his sickness, and Hans said that he travelled hard on the journey, and while they were building houses he did not do any work in the cold, and that did not do him any good. I saw him after he died; I saw Mr. Morton washing him before he was laid out, and then I made a coffin for him. Hays was the one who told me that the captain was dead. It was early in the morning when I was ordered to make his coffin; I made the coffin as quickly as I conveniently could, and he was afterwards put into it and buried on the shore.

After he died Captain Buddington took command. He stated that he should go south ; that he should return home. He stated that his orders were to return home as soon as convenient in case of an accident of that kind. I heard Mauch talking with Hays. He was something of an apothecary and chemist. He had studied chemistry, and I heard him and Hays in a discussion in the forecastle. He was telling Hays that the alcohol that they burned out on the trip had tartar-emetic in it, and that the fumes of it acted as poison when burned. He said he thought that that hurt Captain Hall, I do not know, however, whether there was anything in that or not. Hearing him speak of that, I asked him particularly then what he was talking about, and he told me the same thing. He told me he thought it had a great deal of effect on Captain Hall's health. We staid there during the winter. Nothing particularly was done, except carrying on the scientific observations and making ourselves comfortable. I was engaged in making sleds and mending furniture and working on the observatories, &c., always having enough to do to keep me in exercise.

Nothing special, however, was done during the winter. The scientific department carried on their operations regularly. Mr. Meyers and the doctor were very energetic, I believe, in taking their observations regularly ; though I heard them laugh at the doctor about his getting lost in going over. They had a telegraphic wire afterwards run from the observatory to the ship. I do not know how he got lost. Some state that he was under the influence of liquor; but I cannot say that he was. I merely heard that ; and I do not know as it is proper in me to mention it; for I must say that I never saw anybody that I know of under the influence of liquor, with the exception when they had nothing in the world to do, and those were very active energetic men. One of them was Mr. Chester. I do not think that liquor ever prevented him from doing his duty. I never saw anybody on board the ship so drunk that they could not do their duty. I never saw anybody under the influence of liquor when anything was to be done, that I know of. There might have been and I not have known it. When the spring opened, when the sun first arose, my first business was to go out and take a survey of the vessel and the way she lay. I found she lay very much cramped up. Her bows were on the tongue, on the berg. But what made that was a disaster that happened in the first winter; that is, she broke out in the winter. It was a terrible gale, and we were banked around at the time. The awning was down, and the first we knew the vessel was in motion. We had had ten or fifteen feet of snow banked around her, up to the rail, and the awnings over her. The first thing we knew, the vessel was in motion, and the bank disappeared from around it. There was no possible chance to find out what position we laid in, where we were going, or anything of the kind. There was no light; we only felt we were in motion and under cover, just like as if we were confined below, under the hatches, until Captain Buddington ordered me to cut up some junk and put kerosene on it and make some torches, which I did. After I got the torches made, and lighted them, and opened the port-hole, and stuck the torches out to give light, we could see the iceberg within grappling distance. Billy Lindeman volunteered to go on the berg. He was the man who acted as my mate part of the time. He volunteered to go out and make fast the lines. He cut his foot-holds with the hatchet, and made fast grappling ice-anchors to the berg; and in the morning, or at least as soon as the storm had cleared away, and we got the light of the snow, we saw around us and ascertained the position we were in. It was right on the tongue of the berg. We secured the ship there. We could have no idea of our position until the spring opened. It was fortunate that we got where we did.

When the spring opened we found that the Polaris laid on her bow on a tongue of the berg, and the way she was constructed her rudder-post and stern-post were connected by an extension of the keel, forming a large space for the fan to play in. That was completely locked in with the ice. The shore ice was frozen in on that, and then the berg and pack from the outside forced it up against her, and was lifting her, and the shore ice was holding her down. The berg on the outside pack was forced in against her and there was ice set on the port side right on the port gangway, and that made her position very much cramped. As quick as it came light enough to see to do anything we had all hands turned to and cut a channel around under her counter, and on the star-board side to free her from the pack and from the berg. Still the shore ice that she was frozen into held her in the square where the propeller fan worked between the rudder-post and the stern-post. The condition of the ship was such that she broke on the line of the bottom of the keelson. She broke right through the planking so that it was from an inch and a half to two inches forward extending about eight feet, while there was nothing broken on the starboard side. When the tide was out we got a chance to work on it a little, before we sawed out. We finally got her released. When we started with her she leaked I think three hundred strokes to the hour. That would have been nothing provided we had had fuel enough. There was no impediment to our going. We could free the vessel with our hand pumps, but after that when we started south, when Mr. Chester and Captain Tyson's party came in from the boat journey I had to fix up their boats before they started after they came in. We had made this trial trip up to see if we could not meet them up above. We had not succeeded in doing that and sent for them. We returned to Polaris Bay and took them in. Then we proceeded south about the last of July somewhere I think, or the first of August. When we started south we went, I think, three or four days sometimes in open water. Then we struck into leads and went a considerable time in among the ice, making a good deal of progress. At last we got beset and made fast to the floe, no lead being open we laid there for a long time. This last time we were beset very severely. There was a very strong gale. When we got beset we commenced unloading. I was between the decks at the time. I knew it was pretty severe and believe that the Polaris received very serious damage at that time. I think a piece of her keel was torn off forward. I think this was done by the ice passing under her. She lost a part of her gripe. I do not think she lost any of her stem. There might have been a piece of the keel torn off below the gripe. The piece from the keel to the stem is what we call the gripe. I do not think there was a piece of the stem torn out below the six-foot mark. There was not to my knowledge, at all events, and I made several surveys of the vessel. I think in the first break out that there was a defect in her bottom somewhere. I think when they let go one of the anchors it must have struck the ice and she struck the anchor. If that was so it was on the port side. When we were beset we commenced throwing out everything. At first I was down below, and I got up on deck as soon as possible and went to work passing out the things to the men on the ice. Mr. Chester was receiving from Hayes and me and a sailor by the name of Gustavus. He was a Swede, I think. The greater part of everything was put overboard. They got all my clothing over, and I had nothing and depended on some old clothing that I found I had used up on the voyage.

When the last boat was lowered to some of the men who were on a floe, the vessel separated from the floe. When the vessel separated I had to be at the pumps because they had not gotten the fire up. We were very frugal of the coal and material for fire. We had been working the hand-pumps, but we found that she leaked a great deal more water than before, and we were very quick in getting the fires up. She laid nearly on her beam-ends when the ice slacked away from her. I suppose it was owing to the change of the floe piece and tide. After we did get up fire it was impossible in my mind to steam against the wind and tide, or attempt to reach the floe and the men on the floe. We drifted until morning. In the morning we looked for the men who were left on the ice, but could not see them. I had an idea, whether it was only imagination or not I do not know, but I thought I saw a large number of men on the piece of ice that was nearly like a berg, and a number sufficiently great to indicate that it was our party. I saw no provisions or anything else. They were near enough for me to take in the whole outline of them. It was late at night when we got in at anchor. They were on a piece of ice that was floating. It was moving with the current very rapidly. In the morning when day broke all hands made what little sail we could. The first thing we did was to get up fire. When I thought I saw these men on a piece of ice was in the evening just before dark. During the day we had looked after our comrades, but did not see them. This piece of ice that I saw them on seemed to be going in from Rensselaer Harbor to Littleton Island, in that stream that opened there a strong current. I do not know whether anybody else saw them or not. I mentioned seeing them to different parties, but they did not believe me. In fact I did not want to believe it myself. I thought it was imagination from the way it appeared. This was way up at Littleton Island. I have no reason to believe it was so, because we were lying then at Lifeboat Cove, and if it was them they must have been north of Littleton Island when we got in. They were some three or four miles I think north of where we were then at Lifeboat Cove, drifting as I thought when I saw them. If it had been them and they had no boats, they were in a very bad fix, and we had no boats at all—nothing but the vessel and the fuel all on shore, and we would have had no chance to get at them. It was just before dark of the same night when we went to anchor when I saw them. I reported this fact to Mr. Chester, I think. I do not think I spoke to any one else. I thought myself that it was more likely to be a mirage than a reality. It was from the deck that I saw them as I supposed. The piece of ice on which I supposed I saw them was not the kind of piece where we left them on. The piece we left them on was some miles in extent, with a house on it, and this very piece was a small berg not more than half an acre in extent, some 15 or 20 feet out of the water.

We had very hard work to steam into where we were, and when we got in we found ourselves aground. We built a house at Lifeboat Cove of the spars and bulk-heads of the ship, and we lived there during the winter. Our fuel lasted about half the winter in a small office-stove in the main house and the galley stove. After the fuel was gone, we got fuel from the ship. I made a survey of the ship when they began to get fuel off of her, and handed in a report of her condition to Captain Buddington. I could not tell what her condition was without the ice being cut out of her. She was filled with water, and frozen solid apparently. We commenced to build the boats just the first sun that came. We were only a few days in building the boats. We built three, and a small one we left with the Esquimaux. We started in the boats that were constructed by Mr. Chester and the carpenter to go south somewhere about the 1st of June. After some twenty days' journey south, we were picked up by the Ravenscraig; we went to Lancaster Sound. We went from her on board the Arctic, and came to Dundee. Three of the party were put on board the Intrepid. As to the cold in Polaris Bay and down in Lifeboat Cove, we were not so late in the season, down at Lifeboat Cove, as we were at Polaris Bay. I went in bathing at Polaris Bay, and did not feel uncomfortable until about an hour afterward. There was a storm that came up, and after that it became quite chilly before I got on board again. But while I was in bathing, I did not feel very uncomfortable. I cannot be more particular in my account, unless I had my old log-book here, which I left at Lifeboat Cove. I suppose the log-book that the Tigress found when she went there is mine. That gives the days, dates, and particulars. Sometimes on this expedition I was a little out of my mind. One time I will mention, was a short time while in Polaris Bay.

Examination of Noah Hays.

I am about twenty-five years of age. I was born in 1844, in Henry County, Indiana. Before this, I had been a farmer. That is, I never had any profession. I joined the Polaris expedition, in Washington City, as an ordinary seaman. I rated as seaman, but was coal-passer in the fire-room during the voyage. I sailed from Washington with her, and, afterwards, from New York to New London; from thence to St. John's; from thence to Fiskernaes; from there to Holsteinberg; from thence to Disco; from thence to Upernavik, and thence to Tessiusak, and thence North.

Nothing of importance happened after we left St. John's, before we started north from Tessiusak. After we started to the north from Tessiusak, about the 15th of August, as near as I can remember, I was below a good deal, and did not know as much of what was going on as those did who were on deck. I presume no one knew less than I did. I had no chance to observe anything. Seven hours I was on duty to five off, while the vessel was under steam. The vessel was working along successfully up to the time she was beset, I believe, on the 29th of August. I did not go on deck at all to do any duty. I went from the fire-room to my meals and back. I had only two watches. It was determined that the vessel at the highest point that was reached was in latitude 82°16'. This was when she was beset. She was thought to be higher at the time. I went on deck when she got beset and helped to land provisions on the floe. I saw land on both sides. There was nothing to be seen but ice in front and land on both sides. Ice was all around us. There was a broken pack on one side and a floe on the other. The broken pack was the floating ice and the other side was a solid floe. We made fast to the floe, and when the pressure of the ice was somewhat relieved, we took the provisions on board again, but did not succeed in getting the vessel any farther north. After drifting back perhaps two days, we ran into the harbor. We remained there until the 12th of August, 1872. We went into a bend in the coast there—a little cove behind a grounded iceberg, which Captain Hall called Thank God Harbor. Captain Hall went ashore, and formally took possession of his discovery there. We commenced landing provisions after we put the flag up. We anchored the ship, and began to make snug for the winter. We put up an observatory on the shore as quickly as we could. We banked the ship as quickly as we had snow, and as soon as the ice would bear us up, and housed it over. It was nearly a month, I cannot remember the dates, before we had ice that we could walk ashore on. Captain Hall made a trip to the North on a sledge-journey. He started with a sled for the purpose of selecting a route to see if he could make an overland trip in the spring. He intended merely to prospect, as I understood him. Mr. Chester and the driver, Joe, accompanied him, I believe. I do not think Hans went with him when he first started. After they had gone a little distance they came back for another sled, and then I think Hans went along also. I believe they were gone two weeks. I do not remember anything very distinctly. I kept a sort of a journal, and all these things will be found written down there. During those two weeks nothing happened that I now recollect of any note. We were engaged in banking up the ship with snow. I saw Captain Hall when he came back, and met him at the observatory as he was returning. I asked him about his health, and I had it in my mind up to a little while ago that he said he had been unwell two or three days; but I found on inquiry among the rest of the crew that he told them no such thing, and therefore I must be mistaken about that. He looked very much exhausted to me. I walked back to the ship with him. He went around and spoke to those at work there, and shook hands with them, and went on board, and soon afterward I heard that he had laid down, complaining. After that I never saw him but two or three times until after his death. It was after we went into the cabin that I heard he was sick, but I recollect that he told me that he was unwell. He so appeared to me when I looked at him. I saw him afterward while he was sick two or three times. Only one time I remember of going into the cabin on purpose to see him, but he was in bed, and did not appear to want to talk much. He asked me how I was getting along, and when I told him he said he was glad to see me. I do not recollect what passed between us exactly. That was two or three days after he was first taken sick. He got home about 12 o'clock. We had our dinner at 3 o'clock. I do not know anything about his drinking a cup of coffee, only what I have heard lately. There was coffee on the galley. We always drank coffee for dinner, and we all took coffee shortly after he arrived. I think it was at dinner. It was a little before our dinner time, as well as I can remember, that we had something prepared aft to eat, when we were also called in. This was the coffee on the galley. · I did not feel any bad effects from it, and I did not hear of anybody else that did. Once I went in to see him, and on two or three other occasions I saw him in bed. I do not remember of speaking to him only on one occasion. I just saw him lying there, the same as if he was in a kind of stupor like. I saw him the day before he died; he was lying still and breathing heavily in his bunk. After he died—he was buried on the 11th—I attended the funeral, and all hands were there. Services were read by Mr. Bryan. No one after his death took command formally. I saw Captain Buddington on deck two or three times afterward. He spoke to us in his usual amiable and good-natured style, asking us how we were getting along, &c. We were all engaged on deck at the time, sweeping and cleaning up a little, and feeding the dogs. He was always considered as commander after that. During the winter we had really nothing to do, only to stay about the ship and talk, and take exercise, and feed the dogs. The scientific observations went on during the winter regularly. In March, I believe, Dr. Bessels and Mr. Bryan went southward on a sledge journey. On the 1st of April Mr. Chester, with necessary help, commenced getting ready for a boat journey to the north, as soon as the ice should break away and admit of his starting. During this time there were hunting parties out among the men. I did not go on the boat expedition. Two boats went off on the expedition to the northward. One was commanded by Mr. Chester, and the other by Captain Tyson. Mr. Chester lost his boat the first day, but he soon returned and got another. They were absent from the ship about six weeks, I think. After they had been gone about a month we heard from them, and we tried to get up to where they were. Two men came from Mr. Chester's boat-crew, and told us where they were lying, what their prospects were, and what they wanted. They wanted some more provisions. When the two men came on board from Mr. Chester's party we made another effort to get up there with the ship, the Polaris, but were intercepted by the ice extending from shore to shore, and we could do nothing but to go a short distance above our place of anchorage. We put the men ashore with a small hand-sled, and two bags of bread and some sugar, and such things as that, that they could transport over mountains.

The point we reached the first time, with the ship, was the greatest northing that anybody ever made; that was about the last of August, 1871. We could not see any farther north at any time, because it was thick and foggy; if it had been clear we could only have seen from the deck about fifteen miles, or perhaps less; we could have seen nothing but the horizon, sky, or clouds. I never saw any chance to get any farther north than we did get; I don't suppose there was any chance, except with sledges, and certainly not a very good chance for sledge journeys; there was no land-floe to travel on.

About the 29th of November, I believe, during a heavy gale from the northeast, the ice broke away from us as far in-shore as where the ship was anchored. We commenced drifting away, and fortunately we brought her up against Providence Iceberg. When the gale abated, a day or two after that, where this open water had been left between us and shore, thin ice had frozen over it; we sawed out a place for the ship in that ice, and drew her off from the berg two or three fathoms, as near as I can remember, and also drew her ahead about twice her length, or a little less, leaving the anchor on the bottom, where it was dropped just before bringing against the berg. In a short time after that, there was another violent gale from the southeast; it was quite dark at the time, so we could not see much, but I am pretty positive as to how she happened to be nipped or jammed there. The heavy pack coming in against the berg crowded it farther in against the land, and it came upon the vessel; there was a spur of the iceberg that extended out under the water and it caught on her keel and raised her up. She broke down some of this thin ice, which was not more than 15 inches thick, perhaps—she broke that down two or three fathoms; when the berg stopped moving, she still lay on the spur of it, and was there all winter rising and falling with the tide; that would wrench her. Her stem was not displaced at that time, but was cracked; two or three boards were broken diagonally. The seam of the iron plates with which her prow was sheathed did not run parallel with the boards, and the boards just split diagonally, and opened a crack nearly an inch wide.

We worked three or four days and sawed the vessel out, and as quick as we got loose we made an attempt to steam up north, hoping to overtake the boats; finding we could not do that, we went back and waited till they came aboard. After they came aboard, we finally got out about the 12th August and began to move south. We got along very well for a few hours. When in the vicinity of Franklin Island, in latitude 80° 30′ or 80° 40′, we encountered the ice and were beset there.

The next day it slacked up a little, and we got a few miles farther down, and were again beset, and we tied up to a floe. We remained to that floe, drifting slowly down until the 15th October. We had built a canvas house on the floe, and put some provisions, a stove, and galley on it. On the night of the 15th of October the pack-ice on one side was drifting. The floe appeared to be turning round, which made it seem to us as if the wind was shifting. When it turned around so the wind turned toward the floe, it caught the pack on the other side, drove it away, and left open water there on one side of the ship. The floe kept turning, and in a short time it had turned around so that the wind was coming again over this water, and soon the pack-ice came down on her and nipped the ship severely, threatening to destroy her; but she raised out of the water and was keeled over toward the floe. We then commenced putting everything out on the ice. It was dark and snowing very hard, and blowing a gale then. It was about 9 or 10 o'clock in the evening when she was first nipped. So far as the most of us understood, we thought we should all have to abandon her. About midnight we got off the ice all we could use during that winter; the deck was cleared. We always kept an abundance on deck, in case of accident. After the work somewhat slackened, Captain Buddington gave orders to get everything back from the edge of the floe as quick as possible. We that had been working on deck were then going over the side to help, but the floe had been turning by the force of the wind, so that it left open water on the other side of the ship; and then the edge of the floe had been crushed by the ship coming up and falling down, and had left about 10 feet of ground-up ice between the ship and the floe to which she was tied, so that we could not get on the ice to assist those that were on it in moving the things back. While they were getting out another line and examining the ice-anchors, the ship suddenly broke away. I do not know whether one of the hawsers parted or not; I believe the forward one did. One of her anchors broke out of its place; we hauled that in with the stern line. We then drifted away, and I did not see anything more, as it was dark. We drifted away very quickly. The wind then began to abate. The ship was leaking badly, and, I thought, faster than she had been. We went right to the pumps, made a fire under the small boiler, and got hot water, and thawed out the large pumps on deck, and hastened the raising of the steam. During that time the little pump had been worked all the time for several days preceding, and as quick as we got the deck-pump thawed out so we could use it, we soon cleared her of water. Before that, it threatened to extinguish the fire before we could get up steam. After that we pumped with the small donkey-pump, and we went below and rested in our rooms. It was then getting light again; it was three or four o'clock in the morning. The first thing in the morning we commenced to take down the fore-sail to make coal-bags, knowing we would have to leave the ship. We could not use the pump on deck, as it was so cold, and we had neither force enough to use the pumps nor coal enough to make steam. So we fixed to leave the ship the first opportunity, whether to go on the ice or on shore. We had not force enough to man the deck-pumps and have reliefs.

When morning came I do not know that we looked for our comrades right away. I thought of them all the time, but our attention was drawn to the shore. We soon discovered that we were near Life Boat Cove—the captain and others knowing the position of Littleton Island, and hoped to get the vessel on shore there. I remember that I was that morning at the wheel, and there was a marine glass lying there. I took that, and scanned the horizon two or three times to the southward, to Littleton Island, and to the shore until it brought the other shore within the field of the glass, but I could see nothing of our comrades. Others looked also. Mr. Chester was at the mast-head once or twice, I believe, but he was on deck most of the time. No one was at the mast-head continually. Henry Hobby was at the mast-head later in the day. He said he saw something on the ice, and came down and pointed it out to me. It was a great distance off. I could not tell whether it was a group of men, or what; I thought, though, that I saw something. But afterward the general opinion seemed to be that it was what they called black ice. There is no such thing in fact; but pieces of ice being reared up, it leaves a cavity that, when a long distance off, seems black. It is really a hole in the ice. We thought it might be that. That object was toward the western coast—diagonally across the channel, to the southwest. We could not have got to them, if what we saw had been our comrades. I should think that was fifteen miles from us; it was nearly as far as we could see anything from the deck—nearly to the horizon.

We finally got into land. About 4 o'clock in the evening, I think, we commenced to moor the vessel to the grounded hummocks ashore there. About twelve hours afterward, 4 o'clock in the morning, we commenced to work her in. Fortunately it was high tide when we got on the beach, and we were enabled to run her up so high that she would not sink, but when the tide fell, about 10 or 11 o'clock in the evening, she laid over on her side and took the ground. She would right up each high water and fall again at low water.

Next morning we commenced taking things ashore—timber, planks for a house, provisions, &c. We got everything from between decks as quickly as possible, so as to save all the coal we possibly could, and took everything we could, and let her fall with the water. She was tied well to the hummocks, but gradually through the winter she was crowded off. Every tide the water would come up between the crevices in the ice, and the ice would gradually swell a few inches. No amount of lines would have held her, and before spring the lines were all parted.

Two or three days before we left in the boats she had been blown out of her place, and was two or three times her length below, beating on the rocks, where I thought she would go to pieces, unless wind and high tide would happen to carry her out and sink her.

At this place we built a house and spent the winter. Mr. Bryan made a sledge journey from there to Renssalaer Harbor for the purpose of making some observations for time, and also one to Port Foulke, below there where Hayes had been. Dr. Bessels, I think, went to a glacier near to Port Foulke, known as My Brother John's Glacier, I think. He also attempted to go north. All I know about that is what he told me. He said he wanted to go north as far as possible, and was going to get a good team, drivers, and provisions, and get one of our men, Henry, to accompany him. He said he would make a confidant of me in regard to the enterprise of going up north. I don't think he ever got more than 15 or 20 miles from the vessel. He was gone about a day and a half. When he returned he told us he had crossed the channel and had been a little over a degree above the position of the house; but 1 don't think it was possible by any means. He never notified the captain of his intention. He told Captain Buddington he was going inland to examine a glacier. He returned and said the ice was in such a condition that he could not make the journey he proposed, and said he was forced to abandon the idea. That is the only sledge journey I know of, except those two by Mr. Bryan.

We commenced very early to make the boats, working a little along as the weather would permit. I believe we commenced taking the ceiling out of the cabin in March. Mr. Chester, Mr. Coffin, our carpenter, and one of the firemen, John Booth, worked at it exclusively. The rest of us did nothing with the boats, except get wood from the vessel and ice for water. They were working at the boats at favorable hours from about the 1st of April to the time we started.

We started to go south about the 1st of June with two boats. Captain Buddington commanded one and Mr. Chester the other. We were in those boats working south about three weeks, and got to about fifteen or twenty miles southeast of Cape York, where we discovered the whaler Ravenscraig fast to the ice. We were taken on board of her in a few hours after we first sighted her. She got out from the ice about the 1st of July; I do not remember the date exactly. We went in her across to Lancaster Sound. Fifty or one hundred miles up the sound we spoke the Arctic, from Dundee, and I and some others were put aboard of her. When the Arctic was full and ready to sail for home, we endeavored to get the rest of the crew on board, and spoke the Ravenscraig, and got those on board of her. We also signaled to the Intrepid, but she apparently did not see the signal. She was eight or ten miles eastward, and soon she commenced steaming in another direction. There was considerable ice between the Arctic and the Intrepid, and we could not get to her conveniently. Three of our comrades were left on board the Intrepid—Mr. Bryan, Mr. Booth, and Joseph Mauch. We arrived at Dundee, and finally came home in the City of Antwerp.

The discipline on board the Polaris was very good while Captain Hall lived. After he died I never noticed anything like disobedience, not in the slightest degree. I never heard any complaint made or any objections offered to the commander by any one when ordered to do anything. Still, I think—if you wish me to give any such thing as an opinion—I think it was lax. I think the men did what they were ordered to from principle, and not from necessity at all—not from what they considered necessity, by any means. I consider that, in that respect, we had excellent men.

As to hearing anybody, after Captain Hall's death, say that he was relieved at his death, I cannot remember the exact words; but one day I was over at the observatory with Dr. Bessels. I was there a good part of the time about that time in the winter. He appeared to be very light-hearted, and said that it was the best thing that could happen for the expedition; I think those were the words he used. I do not remember that I heard anybody else say anything of that kind.

Question. Did you ever have any reason to suppose that Captain Hall died anything but a natural death?

Answer. I do not know what it takes to constitute a reason. I never knew anything that would justify any such conclusion. As to what Dr. Bessels said at the observatory, I do not know that those are his words, but it was something to that effect. That was the impression on my mind. I know the next day he was laughing when he mentioned it. I was much hurt at the time, and told him I wished he would select somebody else as an auditor if he had any such thing to say. I was at that time over at the observatory rendering some assistance in the observations, but was not regularly detailed. I have seen Captain Buddington when I believed him to be intoxicated; not very frequently. I never saw him so that he could not do his duty; but I have seen him when I believed him to be under the influence of liquor.

The mean temperature of the winter at Polaris Bay was, I believe, about 20° Fahrenheit below zero. Our minimum temperature, if I remember rightly, was 49°. But then there was some little difference of opinion. I think it was 49° when I observed it myself and recorded it; but I believe Mr. Meyer and others, who would be better authority, thought it was 53°. March was the coldest month. The temperature was not so low as we expected to find it, generally, but still I believe the temperature was lower there than any place south of there. At Lifeboat Cove the mean temperature was a little below that of Thank God Harbor, I believe. I have not looked over the observations so as to determine, but that is my opinion. The summer was not warmer at Thank God Harbor. I have observed closely one thing, and that is, I never passed a day in the arctic regions but what I have seen salt-water ice at some time during the day that had been made during that day. That freezes at a temperature of 28°. There was a large plain right abreast of where we were anchored, and the snow went off of that in June, I believe. The sun pouring right down incessantly on that twenty-four hours a day would cause warm air to come off of that occasionally, which would make the thermometer run up to 40° or 50°. Even then on the shady side there were places, when the sun got around toward the north some 5° or 10° below the horizon, it would be freezing at the top of the water on the shady side of the vessel. It is almost impossible to have thermometers properly protected in the summer time. There is always one part of the day when it is exposed, if not to the direct rays of the sun, to the current of warm air heated by the sun.

There was some vegetation up there—a little moss, several light plants, flowers of moderately brilliant colors, and a little grass. There is not much soil there, or there would have been more vegetation.

The character of the shore was rocky; I think it was limestone, but I know very little about geology. The beach was a shingle beach. The bottom was rocky, with stiff clay between the layers of rock.

There were musk-oxen, foxes, hares, lemings. I saw some wolf-tracks, but no wolves. There were one or two owls seen, too, and ducks and geese. I did not see any auks, but I believe they are there. They also had there what were called ivory-gulls, and another species of gulls that I do not know the name of, partridges, ptarmigan, snipes, and turn-stones, and one or two kinds of plover.

At spring-tide once or twice we had as much of a rise and fall as 7 or 8 feet. It was generally about 6½ feet. At neap-tide it was from 1½ to 3 feet high, as the wind was favorable or unfavorable. The average would be about 4½ or 5 feet, I suppose.

[Diary produced.] That is my diary; it was kept by myself, in my my own handwriting. This one commences on the 15th of October. I kept one before that, but not regularly. I think that diary is here; I gave it to Mr. Chester at Dundee. My position and circumstances were such that I had but little chance to find out anything worthy of note; and the only wonder is that I had a chance to keep any diary at all. These diaries were kept by myself, day by day, as the events occurred, and they will give a more particular statement than I can recollect the details of now.

Question. Is there anything else, to which your attention has not been called or which is not set down in your diary, that you would like to say?

Answer. I am much obliged to you. There is nothing occurs to my mind now that has not been mentioned that I wish to say.

Walter Frederick Campbell examined.

I will be twenty-one years old next Christmas. I was born in Glasgow, Scotland. I have lived in this country seven years. I shipped on board the Polaris as fireman at Washington. I sailed on board that ship from Washington to New York, and from there to New London; from New London to St. John's; from St. John's to Fiskernaes; from Fiskernaes to Holsteinberg; thence to Disco; thence to Upernavik; thence to Tessiusak, and thence north.

Nothing of importance happened between the time we left Washington and the time we reached Tessiusak, the last point on the Greenland coast.

I was engaged principally in the fire-room below, and had not so much chance for observation as those on deck. After we left Tessiusak, we proceeded north for some distance, then crossed over to the west coast, and then skirted up the west coast to the north. We stopped once, and Captain Hall went ashore to see if he could find a place for a depot for provisions, for winter-quarters, if we should find it necessary to come in there. After that we went up through Smith's Sound, through Kennedy's Channel; sailed past Cape Constitution on the right, through what was formerly called Kane's Open Polar Sea, and found it to have land on both sides. We found quite a wide expanse of water between Lady Franklin's Bay and the inlet afterwards called the Southern Fiord; which expanse, after Captain Hall's death, we called Hall's Basin.

We found an opening still to the north above Lady Franklin's Bay, consisting of a channel about twenty-five or thirty miles wide, with land visible on both sides, which Captain Hall called Robeson Channel. We went up this channel until we reached our highest point, in what Captain Hall called 82° 26' north latitude, but which was afterward found by observation to be 82° 16'. We were up in this channel two or three days. At one time Captain Hall tried to make a harbor on the east coast, at a place which he afterward called Repulse Harbor. Afterward in trying to get across to the west coast we got beset in the ice. I don't recollect how many days we were beset there. We put provisions out on the ice, and kept shifting them about, taking them on board and putting them on the ice again, as the danger appeared to be more or less imminent. Afterward, with wind from the northeast, we drifted for two days farther south, when the ice slackened, and we made in to the east shore in a small cove in the lee of a stranded iceberg, which Captain Hall called Providence Iceberg, calling the harbor Thank God Harbor. We lay there about three days, and when the ice got thick enough we got provisions ashore. We then made the ship fast to the berg, and some time after Captain Hall's death we broke out with a northeast gale. That night we got our ice anchors fast again, and that was our rescue. After we got fairly into winter quarters Captain Hall made a sledge-journey to the north, and was gone somewhere about two or three weeks. On his return from the sledge-journey I was the first man that met him. I met him above the observatory, on shore. I asked him if he was fatigued after his journey. He said "no, he was pretty tired, but quite well in health," and came to the ship and made a hearty greeting to us all. I walked down to the ship with him. He looked tired, and that is just the reason I asked the question if he was fatigued. He looked tired and worn out. He said he was a little tired, but in good health. He shook hands with all the men, who were at that time banking up the ship, and afterward welcomed us into the cabin. I was second steward that winter. I was helping John Herron, who was steward. Captain Hall came into the upper cabin. I went down into the under cabin. I heard the captain ask the steward if he had any coffee ready, at least the steward first asked him if he would like to have a little coffee, and he said if all hands would have coffee he would be glad to join them. I really forget whether they did have coffee or not, but I believe Captain Hall and all hands did; indeed I am quite sure they did; and afterward, that night, he took sick. The steward got the coffee from the galley. It was made purposely for the captain. I could not say whether all hands had coffee or not, but several of them had, I know. I didn't see the steward get it from the cook, and I didn't see the coffee prepared. The coffee was had in the upper cabin and in the lower cabin. It was taken up into the upper cabin in a kettle, and afterward the same coffee was taken down to the lower cabin in the same kettle. I think it was the steward carried it. Afterward I had to wash up the dishes, and then I went forward and retired. I lived forward. I don't think all in the cabin did have coffee, but I am not sure, and didn't pay much attention to it, but I know several of them did have coffee, for I washed the dishes. Some of the men were playing chess, some sewing, some washing, some reading and talking. Noah Hayes plays chess, and Mr. Coffin, the carpenter, and I believe Kruger, and there was checkers, too. The next morning after Captain Hall's return was the first I heard of him being sick. I didn't hear of it until next morning because I went to bed as soon as I got my work done. I saw his face and head several times during his sickness, but didn't speak to him. It was my duty to go into the cabin in the morning and sweep it out. Mr. Schumann, the engineer, the doctor, Mr. Bryan, and Mr. Meyer lived in the cabin, and there was John Herron and the cook in the upper cabin. I believe that was all at that time because the berth above Captain Hall was not occupied. I never spoke to the captain again, and only saw him occasionally during his sickness when I went into the cabin to sweep it out. I know nothing about the captain's sickness only as I have heard the talking among the men. Some said that he had had a sun-stroke some years before he went up there, and it had affected him on his sledge-journey. Another thing I heard was that some of the men asked Dr. Bessels what he thought, and the doctor told them that he would never get over it. This was when he was first sick. I am not quite sure what man said that, but I believe it was Herman Simmons, if I am not mistaken. After the captain died he was buried on shore.

Question. Did you ever have any reason to think that he died anything but a natural death?

Answer. Well, sir, I have got no idea about it at all, and I could not have anything to say on the subject. I don't know of anything that would lead me to any other belief than that he died a natural death. I do not know anything that would give me ground to suppose that he died anything but a natural death. After Captain Hall's death the first report that I heard was that Dr. Bessels was to have command of the ship. Then I heard that Captain Hall had turned the ship over to Captain Buddington. Three or four weeks after the captain died John Herron didn't have so much to do, as everything was cleaned up for the winter, and then I had to work after that in the engine-room all winter. Nothing happened of importance after that except the blow which broke us out from our anchorage and drifted us out against the berg. The ship rested on the heel of the berg during the whole winter, rising and falling with the tide. Every tide she would rise, and when it went down she would just lay right over. The scientific operations went on all winter. When spring came Mr. Chester and Mr. Tyson made two boats ready. They were about a month in preparing to go north. Mr. Bryan and Dr. Bessels had been on a sledge expedition toward the south. That was before the boats went. The boats were gone, I believe, exactly a month, and while they were gone we sawed the ship out. While they were gone we made three attempts to get north; but were not able to do it on account of the ice being so thick in the straits. The ship was then making a great deal of water and we were obliged to pump her regularly. I was attending to the donkey-pump, and kept it going about twenty or twenty-five minutes out of the hour sometimes. She didn't seem to leak so much afterward. She kind of filled up with sand after we returned from the north and got back to winter-quarters; the sand and clay together entered the crack and stopped it up.

Finally, on the 12th of August we started southward. We steamed along until we entered into Smith's Sound, where we got beset in the ice again and drifted farther south. Sometimes we would see a little crack of open water. There was open water in toward the west shore, all the time. That looks as if, could we have got there, we could have got down. We finally tied fast to a floe, and floated on it two months or more. We built a house on that floe; put some provisions on it and staid there till the night of the 15th of October. At that time I was below and I felt a kind of motion in the ship that I thought kind of curious, and I came up on deck. Just then the crack was opened, and I went down to report to the chief engineer. On going down I met him in the engine-room and he sent me down to steady up the boiler and keep it from falling; and after we had steadied it up we ran afoul of this berg and the ship canted and went over; and it was as much as we could do to get back out of the engine-room; but we did get back in time to assist in getting the provisions over. The plates in the fire-room lifted it and there was great difficulty in getting over; but we finally got out. We assisted in getting the provisions out. Afterward there were men— I can't say who, sent on the ice to pick up the provisions on the edge of the floe and take them to the house. There were several, and I believe Henry Hobby was one, standing on the gangway. The ice was shifting around about the provisions and Mr. Chester sent me to the pumps to pump by hand; and after that I couldn't see much of the proceedings. We were getting all the stuff out of the cabin and putting it over the side. The ship seemed to make water rapidly after that and the water gained so fast that it was as much as I could do to get steam on. It was very dark and it was blowing and there was a heavy snow-drift. Her moorings did not hold her, and she drifted off in the gale to the northeast, or somewhere about that direction. After we got adrift we had hard work to keep her afloat; at such times as we gained on the water, we got her clear. All hands worked at the deck pumps till we got fires under the boilers. They were working at both deck pumps to keep the water out of the fires, and finally succeeded in getting up steam, and we then pumped by steam.

Next morning we were surprised to find ourselves near Life-Boat Cove; the storm had then moderated considerably and cleared up. As soon as it came daylight we made fast to some little pieces of hummock. As soon as it was light enough Mr. Chester went aloft to the masthead to see if he could get any tokens of the party that broke adrift from us. He could see provisions, but no boats or human beings. In fact, I went up myself, in a little space of time, and I could see nothing but a few boxes and stuff on a piece of ice. I know I saw some provisions on a piece of ice, but I could see no tokens of any human beings. I believe it was about 10 o'clock in the morning when I went up to the masthead; the weather had cleared up, and it was a very nice day and quite calm. We worked very hard that day trying to get the ship to the beach; she was making water all the time, and we had to keep the pumps going most of the time. We could have got up sufficient steam to work the pumps, but our fuel was scarce and we were looking out for that. We wanted to save all the coal we could. We kept up just enough steam to keep her dry, and I believe it took us about three days to get all the stuff off and to let her fill in. We landed them, and Mr. Chester erected a house on shore. The Esquimaux from the nearest settlement came the second day after we got there to visit us. They helped us land the provisions, and one family from the west side staid with us all winter.

Dr. Bessels and Mr. Bryan were at work all winter at their scientific operations. Dr. Bessels made one or two attempts to go North in the spring, and made another sledge journey to the South. Mr. Bryan also went North to Rensselaer Harbor.

When spring came we built two boats, commencing about the 1st of April. On the 3d of June we started south. We were on the boats about twenty-one days. When we got about fifteen or twenty miles south of Cape York we were picked up by the whaler Ravenscraig, and went with her to Lancaster Sound, then we were transferred to the Arctic, and went with her to Dundee, and came thence to the United States.

I did not keep any diary while the vessel was under way, my duties kept me below, so that I did not have much chance for observation.

Q. Is there anything that you can think of and wish to say to which your attention has not been called?

A. No, sir; I do not think of anything.

Q. How was the discipline on board the ship while Captain Hall lived?

A. Everything was orderly, as far as I knew. I tried to do my duty, and everybody else did the same; in fact I did more than my duty, I did all I could.

We had a cat on board that we took with us from Washington. A little soldier boy had it on board the Polaris at the Washington Navy yard before we sailed and he gave it to me. We took the cat with us, and he staid with us both winters in the ship, and finally ran away from us at Hakluyt Island as we came down in our boats this last spring. The Esquimaux at Life-Boat Cove had never seen a cat before and were very much interested in it. They gave it the name we called it by, "Tommy." They have a name for it in the Esquimaux language, though they have not the animal itself. I do not know the name.

WASHINGTON, D. C.,
December 24, 1873.

At 12 o'clock m. Hon. George M. Robeson. Secretary of the Navy, Admiral Reynolds, Professor S. F. Baird, and Captain Howgate of the Signal Service, assembled at the Navy Department for the purpose of taking the statements of the last three of the survivors rescued from the steamer Polaris, and who arrived in New York on the 6th day of November, 1873.

Examination of Richard W. D. Bryan.

I was born in 1849 in the State of New York, and am twenty-four years of age. My last place of residence was Westchester, Pa. I joined the Polaris expedition as astronomer. I joined the Polaris up at Disco Island, in the harbor of Godhavn. I went out in the Congress. I could not tell you exactly the date I got on board the Polaris, but I think it was the 13th or 14th of August. We left Disco on the 17th of August, I think, and ran up the coast, keeping within sight of it all the time and stopped at Upernavik. We stopped there for a couple of days. Then we left there and touched at a little place that they called Kingituk. We merely sent a boat ashore there. We did not anchor the vessel, but only staid there about an hour. We then proceeded up to a place called Tessiusak, the last Danish settlement. We remained there until the 24th of August. On the afternoon of the 24th we started again, and kept along the coast until we came to the entrance of Melville Bay. Then we struck a course for Cape York, which we sighted the next day, I think. After passing Cape York we kept up along the coast, passing between Wolstenholm and Saunders's Islands. We were then compelled to keep more to the westward on account of the ice, and went on the outside of Hakluyt Island. Then we were enabled to go more to the eastward. We kept quite close to the east cost when we passed Cape Alexander. When we got up to Cairn Point, however, we were driven over to the westward to find an opening through Smith's Sound. Then we took very nearly a straight course for Cape Frazer, at the entrance of Kennedy Channel. I cannot remember the date that we arrived opposite that, but at any rate we stopped there, and Captain Hall went on shore to look at a bay there, and to see whether it would answer for a harbor in case the vessel should be stopped by ice. It was on the western shore of Kennedy Channel. He came back and said the bay was too shallow to anchor the vessel in. Then we ran up quite close to the west coast of Kennedy channel. We were first stopped by ice on the 29th of August, I think, when we got our latitude. That was the only latitude, I think, that we got after passing Cape York. The latitude was 81° 20'. We only remained there part of that day. In the evening we started up Robeson Channel, and gained our highest latitude on the 31st of August. Then after we gained our highest latitude it was decided that we could get no farther north—at any rate, on that side of the coast, and it was decided to try to go to the other side to find a lead up along that side; and if we were not able to do that, then we intended to return. At that time we were quite close to the east side, when we gained our highest northern latitude much closer to the east side than to the west side. It was found that we could not get any farther on that side, and then it was decided to endeavor to penetrate the ice and get to the west side if possible, we supposing that we might find a lead there that would carry us north. We endeavored to do that, and in doing that we got beset.

I think it was about noon that we reached the highest latitude. We tried to get over toward the western coast, but on our way over we got beset; it was decided we could not get up any farther on the east side by those who had charge of the vessel. I did not know much about it myself. I did not go off the deck at that time. About 12 o'clock there was a consultation called as to what course should be pursued. I believe that the consultation was called because Captain Buddington had told Captain Hall that they had gone as far as they could. I was not present at that consultation, and they did not ask my advice in regard to the matter. I learned afterward that the result of the consultation was that they would endeavor to get to the west side in order to find a passage. It was determined that if they did not succeed in getting a channel up along the west side, then they would return to the east side, and run into a harbor that had been seen on that side. It was in trying to go across that we got beset. While we were up there at that highest point, we were all the time looking out. A good part of the time it was very foggy, and it was snowing. There were drifting snow and snow-squalls, so that it was only at times that we could see the land. For a short time, however, we had very clear weather, and then I could see the land on the east side, which seemed to end in a point. I saw, also, the land on the west side. The land on the east side I followed up a short distance with my eye, with the aid of a glass from the deck. I did not go up aloft. Far ahead we saw what the sailors call a water-sky. A good many thought it was land. I could not see any indications of there being land there. All around us was very heavy ice, and it was moving very rapidly down the channel, and, as I say, there was what the sailors call a water-sky. Right around the vessel there was quite a space between the different floes, so that I was, personally, very much provoked that they did not go up farther; but I have since learned that a person from the deck of a vessel cannot form a very good judgment in regard to ice. I learned this from experience that I had on board the whalers. On board the whalers I looked at the ice from the deck, and then went up to the mast-head and looked at it through glasses, and I found that a person could not form a correct judgment at all from the deck. It looked to me at the time, however, as if they might have gone on. I suppose, even now, that they could have gone on for, perhaps, half a mile, but I am very well satisfied that they could not have gone any farther. As I remarked, however, I was at that time of a different opinion.

As I stated, we got beset in the ice and drifted down. We drifted for I think a little over three days. On the 4th the ice opened somewhat, and we got a chance to get the vessel out. We steamed right into the east coast, and dropped anchor there. It was on the midnight of the 4th that Captain Hall went on shore for the first time. This place where we anchored could hardly be called a bay. It was part of a large bay that is formed there that Captain Hall afterwards named Polaris Bay. The particular place where we dropped our anchor could not be called a bay, however, nor was there any particular harbor there, but it was out of the current because it was under the lee of the cape, at the entrance to Robeson Channel, which Captain Hall called Cape Lupton. The current for this reason swept the ice clear of us, and at the same time we were on the inside of a large iceberg, which it was thought would protect us from the pack coming up before the southwest wind. Captain Hall went ashore there the first night at midnight. I think he there unrolled the stars and stripes; at any rate, he told me when he got back that he had taken possession of the land in the name of the United States. He said he went there for that purpose.

It was decided to remain there, and to make our winter quarters there. We then commenced work, landing our stores and provisions on the shore, and otherwise to prepare for winter. We also put up an observatory. Everybody was engaged in doing something. I cannot tell exactly what day, but later, a sledge party was started out by Captain Hall, which consisted of Mr. Chester, the first mate, Dr. Bessels, and two Esquimaux. They were sent out to try to find some musk-cattle, traces of which had been seen by the Esquimaux, and also for the purpose of ascertaining whether there was a feasible overland journey to the north.

They were gone six or seven days, and brought back one musk-ox. On the 10th of October, I think it was, Captain Hall himself started on a sledge journey with Mr. Chester and the two Esquimaux. On the 24th of October I think he returned. He was gone at any rate two weeks. During his absence the observatory was put up, and the ship arranged for winter quarters. Part of the awning was placed over the vessel, and the vessel banked a little with snow. Observations were commenced. There was some little surveying done. I cannot now think of anything else.

Captain Hall found a bay which he called Newman's Bay, after the Rev. Dr. Newman, and followed that out to where it empties into Robeson Channel, and called the two headlands—the one to the south Summer Headland, and the one to the north Cape Brevoort. This was but little above latitude 82°, I believe. This was the farthest point Captain Hall reached.

Captain Hall crossed the bay, and had one of his encampments right under Cape Brevoort. Then, finding that he could not continue further with his sledge upon the ice, he took a walk one day over the hills. I do not know how far he went, but the copy of his journal, in which I presume that was noted, was brought down, I believe. He came back and reached the ship on the 24th of October, and was at once taken sick. He remained ill for two weeks; conscious part of the time, apparently, but most of the time quite delirious. On the 8th of November he died. I saw him when he came back. I was on the deck of the vessel. I saw him on the ice coming up with the sledges, and then I spoke to him. I do not remember whether I went over on the side of the vessel to speak to him, or waited until he came on board; but I remember that I spoke to him at the time, and remarked that he was looking very well. I think he said as usual, "I am very well, I thank you," or something of the kind. I did not notice anything particular about his remarks. He did not say he was not well, but Mr. Chester told me that he thought something was the matter with him on his sledge journey, that he was not quite as active as he would expect him to be. Captain Hall mentioned this fact to Mr. Chester, that he was more inclined to ride on sledges than usual, and he mentioned the fact to Mr. Chester as something unusual; that ordinarily he was able to run along with the sledges without riding on them, except once in a while; but he was compelled to ride on this journey more than was customary with him. Mr. Chester told me that during Captain Hall's sickness, I think. I will not, however, be positive about that. It is a long time ago, and it might have been either during his sickness or after his death that he told me.

I lived in the lower forward berth on the port-side. I lived in the same cabin with Captain Hall; but there are two berths, one above the other; I lived in the lower forward one on the port-side, as I have stated. Captain Hall at that time was sleeping in the lower after-berth on the starboard-side in the cabin. He had removed from his little room, and fitted that up for a galley. I saw him and shook hands with him when he came on board the ship, and in a very few minutes, I think, I followed him right into the cabin. I remember Mr. Morton was seeing about getting his wet boots off, and I remember his drinking a cup of coffee. Then he got up to change his shirt, and he said, "I feel sick," or "something is the matter with me," or something of that kind. He made some such remark as that he was very weak. Then Mr. Morton and some one else assisted him into bed in his berth. I did not think he was very sick, not at the time. I thought it was just probable that he had over-exerted himself. I did not think he was at all sick then. It was a very few minutes after he got into the cabin. He had just stepped on deck and spoke to a few of us, and then walked right into the cabin.

Question. Was this within half an hour of his coming into the cabin or coming on board the vessel?

Answer. Yes; I think it would be safe to say it was within that time.

Question. Did he then take the coffee?

Answer. Yes; I think I saw him take the coffee, and almost immediately afterward——

Question. Within five minutes afterward?

Answer. I do not know about that, because he might have given the cup back, and he might have spoken a little while, and my attention being turned off to something else, I could not see whether he took it or not; but I associated the two facts in my mind, that just as soon as he took the coffee he complained of feeling sick and went to bed. It might have been more or less of an interval; I could not tell you exactly how long.

Question. When you say you thought he had over-exerted himself, did he seem to be weak? Is that the idea you wish to convey?

Answer. Yes; he seemed to have something the matter with his head I thought. That is what I thought when he was first taken, and I have an indistinct remembrance that he threw up after he got to bed; but I won't be quite positive about that. I thought he was just fainting, or dizzy, or something of that kind.

Question. The impression made upon you by his conduct then was, that he had something the matter with his head, or was rather faint and dizzy?

Answer. Yes; that was my idea at the time. ·I think that shortly after that he threw up. It just occurred to me now; I never thought about it before.

Question. Who brought in the coffee

Answer. I think it was the steward. I could not tell you where it was brought from. I have no doubt that the steward brought it.

Question. Did you take any of that same coffee?

Answer. Not at that time. I might have done so previously or subsequently. I do not know positively about that, but I did not take any coffee at that time.

Question. Do you know whether the coffee was brought in the same identical coffee-pot that was used in the galley?

Answer. No; I do not know anything about that. I had no chance of knowing. I did not go to the galley to find out, and I only saw the steward enter the cabin with a cup of coffee and go out with an empty cup.

Question. Did anything occur to you as a matter of any interest?

Answer. No, sir.

Question. What happened after he went to bed?

Answer. I cannot give you the events in the order in which they occurred. I kept a journal. I just remember that Captain Hall was very delirious at times, and at other times quite rational. That is, he seemed to talk very reasonably about his plans for the future and about himself; but the most of the time he was out of his head. I saw him every day. I slept with him; that is, I slept in the same room with him.

Question. Who took care of him during his sickness?

Answer. There were several. Mr. Morton and Mr. Chester seemed to take the task on themselves more than any one else. During the daytime several of us would stay with him, but during the night Mr. Morton and Mr. Chester were with him. These are the only two that I remember as having set up with him at night. I know of several who offered to, but Mr. Chester and Mr. Morton seemed to take it on themselves.

Question. After you went to bed the first time, did you see anything of him before the next morning? Did you hear anything more about his being sick?

Answer. I could not tell you that. I do not know when I first got the idea that he was really sick. I never supposed he was so sick that he would die until he did die, although Dr. Bessels used to say that if he had another attack he would die. I remember hearing Dr. Bessels saying that, but then I did not believe it.

Question. Who attended him? Did he take any medicine?

Answer. I do not know what medicine he took; however, I remember the doctor once gave him a mustard-bath. He bathed his feet in mustard-water, and then he used to give him hypodermic injections. I know that the doctor at one time wanted to administer a dose of quinine and the captain would not take it. The doctor came to me and wanted me to persuade Captain Hall to take it. I did so, and I saw him prepare the medicine. He had little white crystals, and he heated them in a little glass bowl; heated the water, apparently to dissolve the crystals. That is all I know about any medicine. I only knew that, because I had persuaded Captain Hall to take the injections. It was given in the form of an injection under the skin in his leg. I believe he gave him medicine at other times, but that was the only time I had any knowledge of it.

Question. Did you have any difficulty in persuading the captain to take it?

Answer. No, not very much.

Question. Why did he object?

Answer. He did not like the doctor very much at that time, and he was a little delirious, I think. He thought the doctor was trying to poison him.

Question. Did he ever tell the doctor so?

Answer. O, yes, repeatedly; but then the doctor was not the only person that he accused of murdering him. He is the only one, however, he ever accused of *poisoning* him. He accused nearly all the officers of the vessel at one time or another of trying to murder him, I believe; I have no idea, however, that he was in his right mind when he made those accusations; I did not think so then, and I do not think so now.

Question. Did he ever accuse you?

Answer. No, sir; I do not think he ever accused me, but he did nearly all. I do not remember of his ever accusing me.

Question. How was it about Mauch?

Answer. He had a good deal of confidence in Mauch, but Mauch was not with him very much in his sickness.

Question. Did he ever accuse Joe or Hannah to any one?

Answer. No; I think not. He lingered for two weeks, I think. I think he was taken sick on the 24th of October, and died on the 8th of November. He was better some of the time; he then appeared to be quite rational indeed, and he spoke very well. He had Mauch in the cabin one day, and he was looking over some records of his sledge-journeys, trying to get them fixed up, and discussing his plans for another sledge-journey.

Question. Did he appear to have any misgivings at any time that he would not recover? Did he ever refer to the probability of his not getting well?

Answer. When he was rational?

Question. Yes.

Answer. No, sir; I spoke to him about it once, and told him he might not get well, but he did not seem to think that there was any immediate danger. That was when he was nearest in his right mind. It was pretty difficult to tell when he was in his right mind and when he was not, because sometimes he would get off something very rational, and then he would come out with something that did not sound so well.

Question. Is it your idea then that for almost all the time after he was first taken sick until he died his mind was unsettled?

Answer. Yes, I think so. Occasionally he would appear to be nearly rational, but then he would break out again into saying strange things.

Question. He accused almost everybody, you say, of wanting to murder him. Do you remember any particular instance?

Answer. It is pretty difficult for me to distinguish between what I remember from my own observation and what I remember from hearing others talk, because we have spoken about all these things so much. I think I was in the cabin at the time he accused the cook of having a gun that he was pointing at him from his berth. And then he used to frequently remark to me that the doctor had some infernal-machine there in the berth that emitted some blue vapor. He said he could see the blue vapor coiling all around in the atmosphere, and hanging alongside the edge of the berth; and he would call my attention to it, and ask me if I did not see it. He would say, "Now, it is there crawling along your nose." He said that the doctor had put that machine somewhere, and that he was pumping this blue vapor into his berth, and it was killing him. Then I have a faint recollection that I was in the cabin when he was complaining about a conspiracy that had been formed by the officers. I think he was complaining to Captain Buddington at that time. He thought that Captain Buddington, Mr. Chester, Mr. Morton, (I do not know certainly about Mr. Morton,) and Captain Tyson had joined together to kill him; but that, I suppose, was just his wanderings.

Question. Did Captain Hall exhibit any symptoms of paralysis, as far as you could judge?

Answer. I heard he was paralyzed all on one side, but I never noticed anything of the kind.

Question. Did you discover any difficulty in his articulating distinctly, or in swallowing at any time?

Answer. I never noticed anything of the kind. I did not have much to do with him when he was eating. He was very particular about his eating. For a long time he kept his food under his own charge, and got Mr. Morton to administer it to him. His food he kept locked up under his berth, and took the key to bed with him.

Question. Do you know what he did eat?

Answer. He had crackers, and I think they made him some kind of gruel, or arrow-root, or something of that kind. And then he had in his drawer a bottle of wine, and, I think, a little preserves; but finally, I believe, he intrusted the care of it to Hannah, so that Hannah was the only one who administered any food to him; but I never heard at the time, that I remember of, that there was any inability to swallow on his part.

Question. Do you remember when he died?

Answer. I remember the night he died.

Question. Where were you?

Answer. I was asleep. There was no one up but Mr. Morton. I was called up.

Question. Had he had a second attack just before he died?

Answer. I do not know. I did not understand these attacks at all.

The night before he died, as he went to bed, he appeared very rational indeed. I remember this very distinctly. I was there at the time, and the doctor was putting him to bed. The doctor had gotten him into bed, and was tucking his clothes around him, when the captain said to the doctor, "Doctor, you have been very kind to me, and I am obliged to you." I noticed that particularly, because it was a little different from what he had been saying to the doctor. I think these were the last words that he uttered, because that was just as he was fixed for the night, and then he turned over and went to sleep. Mr. Morton told me that all the evidence that he had that he was dead, was a cessation of breathing. He said that just before he died he had heard him with his regular breathing, and then all of a sudden his breathing ceased, and then it commenced again. I think he said it ceased twice, and then altogether, and then he woke us all up.

Question. Did you notice his breathing at all when he was sick; was it loud breathing?

Answer. I noticed that it was a little louder than usual; a little stronger than a person ordinarily breathes. After he died they prepared him for burial, though I was not present at the time. I did not see him until after he was laid out in the coffin; that is, I did not see the preparation. I just remember of going into the cabin and seeing the coffin on two chairs with Captain Hall's body in it. I believe the carpenter made the coffin. I think he was buried on the 11th. I read the service. I did not read all the burial service; I just read portions of Scripture and offered a prayer. It was what we call the day-time, though it was quite dark. There was, however, considerable twilight. That particular day the sky was very cloudy, and you could only see the glimmering of the twilight through the breaks in the clouds. They were heavy clouds—heavy water-clouds.

Question. After Captain Hall's death and burial, who took command of the expedition?

Answer. At first there was not very much of a change. Whenever there was anything to do, Captain Buddington always had it done. There was not very much for the crew to do, except to clean off decks, and sometimes to go ashore and get some provisions. Captain Buddington would always tell the first mate to have such things done, so that there was not much necessity for any exercise of command. Whenever the crew had a complaint to make, they always came to Captain Buddington and made it, and he tried to have the thing fixed up. I believe Dr. Bessels got up a paper that he called the first consultation between himself and Captain Buddington, which he signed, and I believe Captain Buddington also signed. I do not know exactly at what time that occurred, but the paper did not amount to anything, except the statement in it that they both proposed to do their duty. That is all.

Question. Then Captain Buddington went into command upon the death of Captain Hall?

Answer. Certainly. We all recognized him as commander. He did not get up and say he assumed the command and direction of affairs. There was no formal announcement, but he took command as a matter of course, just as a lieutenant of a company would take command upon the death of the captain, or the mate of a vessel take command if the captain was shot. There was, however, a good deal of talk at first about there being a joint commander, on account of the instructions of Captain Hall. Some contended that it was intended that Dr. Bessels should be joint commander with Captain Buddington. But Captain Hall had left no written instructions to that effect, and, of course, that was no argument at all. Still, that proposition was advanced by some, but only by a few. We continued along in that way. The crew did not have much to do. Observations, however, were kept up. The first thing that disturbed our winter life was a very severe gale on the 21st of November. That was a very strong gale from the northeast. After the gale had blown some time, we heard water dash up against the side of the vessel, and then we knew that we were adrift. We were very much afraid then that we would be driven out into the pack. The cable was played out so as to let the ship swing to her anchor; but after a short time they began to see that the anchor did not take any more cable, and yet the vessel was broadside to the wind. At first they did not know how to explain the phenomenon, but they looked on the lee-side of the vessel and found the vessel was lying right up against the iceberg, and that the iceberg held it from going with the wind. Then they sent two or three men out.

Question. Was this in the night time?

Answer. It might just as well have been night, for we could not see anything. Besides, we had awnings all over the vessel. It was so dark we could not see very much. The men went out on the iceberg and we lighted them up by putting tarred rope in a pan with kerosene oil and setting fire to it. They went outside, and put two or three anchors in the berg, and in that way the ship held fast until the gale blew over. In a couple of days ice was formed around the ship again, and then the ship was drawn off from the berg about 50 feet and about 100 feet farther on, so as to get it more fully under the line of the berg. We continued that way quietly for a few days, when, on the 28th, we had a very strong gale from the southwest, just the other point of the compass, and that gale had the effect of driving the berg, although it was aground, over this 50 feet right up against the vessel and pressing the vessel against the ice on the other side. But this ice happened to be young ice that had been formed since the northeast gale, and it broke, the consequence of which was that it saved the vessel. If that had not broken, of course the vessel would have been crushed. When this berg came in, there was a tongue run under the bow of the vessel. I do not know whether at that time it split the stem, or afterwards; but the result was that the stem was broken. It not only ran this tongue under the vessel, but behind the vessel and on the outside of the berg, it piled the ice up very high, the young ice being broken by the pack on the outside coming in. It piled the ice up so that we could stand on the quarter deck and step over on to the ice. After that gale was over, some few efforts were made to try to get the vessel clear, but we could not get the vessel forward on account of the vessel resting on this tongue of ice. The only way to move the vessel would have been to run her back, and that could not be done on account of this pile of hummocks that was at the stern of the vessel. It would have taken a great deal of time to have got these away if it had been possible to get them away at all with the force we had. Several efforts were made to blow the ice up around with gunpowder, but they proved ineffectual, and it was decided that the only thing we could do was to allow the vessel to remain there, which we did. She remained there all winter. On account of being tilted up against the berg, and the berg remaining aground, and the ice rising and falling with the tide, it was rather uncomfortable on board the vessel, because at times in low tide she was tilted over so that it was very difficult to walk up from the port to the starboard side.

We continued in that way all winter doing nothing very much, except, of course, keeping up the observations all the while. The men, however, were not employed in any work except every morning to fix around the deck a little. This state of things continued until, I think, it was in February, Dr. Bessels prepared a plan of operations for spring work. In these operations he proposed to send three sledge journeys, one to the eastward to endeavor to reach the east coast of Greenland; one to the southward to join Kane's survey with ours, and one, if possible, to cross the straits to get on the west side. Then he afterward proposed to join the last two sledge journeys together, and let the men who went down to join Kane's survey with ours also cross the straits, if it were possible, on the ice. He proposed that these two sledge journeys should start at one and the same time, and that they should be back in time to start with the boats if there was any opportunity in open water in Robeson Channel. Then he went on to detail in this letter the different plans that the boats should pursue, the direction which they should go, and the manner in which they should provide for the vessel meeting them, or something of that kind. Then Captain Buddington in reply to this letter, approved of Dr. Bessels's plan of sledge journeys, and said he would do all in his power to carry them out, but that as regards the boat journey he intended to send it himself. He considered it too early, however, to make any arrangements as to the details of the journey. I suppose he wrote that letter because the instructions gave Dr. Bessels control of the sledge journeys in addition to that of the scientific work. Along in March, I think it was the 27th of March, Dr. Bessels proposed the sledge journey to the South, to join Kane's survey with ours, and I accompanied him. We took two natives with us. We started out with one sled, and one native, but the native found it was too hard work for him alone, so he wanted to go back, and he did go back and brought the other native along with him and another sled; we went down into Kennedy Channel along the east coast a little distance, when we came to open water. We could go no further with the sledges, so we returned to the vessel. We were only gone about a week.

Question. How far did you go, and what did you find out on that journey?

Answer. We found out that Kane's farthest point—Morton's farthest point when with Kane—what is called Cape Constitution, was in a little lower latitude than is represented on the maps.

Question. Did you go down to that point?

Answer. No; we could not reach that point. We went down in sight of it, but we could not reach it on account of the open water—probably the same open water that Mr. Morton saw. It was in the same position at any rate, but we could not reach it. The ice foot gets very narrow in that place, I suppose on account of the strong current tearing it away as fast as it forms; and what little there was of it was piled up with large pieces of ice, so that we could not get a sled over it at all. We walked a long distance over it, but there is a limit to walking expeditions, especially when you carry no food with you, and have to go back to the sled to get something to eat. We went down however until we saw Cape Constitution, which was about 20 miles off, I presume.

Question. Twenty miles to the southward from where you went?

Answer. Yes.

Question. Did you take any observations there?

Answer. I did not exactly take the observations for latitude at the farthest point we reached on the coast, but I took observations for latitude at a little island where we made our encampment, and I made the latitude 81° 5'.

Question. And Cape Constitution is how far south of that?

Answer. I suppose it is about 25 or 30 miles south of that.

Question. About one-half a degree?

Answer. Yes, sir; a little less than that.

Question. How do you know that was Cape Constitution?

Answer. We knew it from the description we had of it. We were not quite sure at that time, however, although Hans who was with us was the same man who was with Morton. He said it was the same place, but still we did not regard that as very reliable, inasmuch as you can hardly expect a man to remember a place that he has only been to once, and that nearly 25 years ago. But coming down in the vessel we drifted very slowly past that, and then we had an opportunity to see it, and we knew it was the same place that we had seen before. As I said, on this journey we were gone about a week and then returned to the vessel.

Question. Did you go to what is called the Southern Fiord?

Answer. Yes, we ran into that some distance—some twenty or twenty-five miles.

Question. Did you find how deep it was?

Answer. No, we could get no end to it. We were stopped by ice-bergs. The icebergs ran right across it. We could not get the sled up at all. On board the vessel not much was done except to get the boats ready to start. Captain Buddington ordered Mr. Chester and Captain Tyson to be ready to start by the 1st of May, I believe, and then to start after that just as soon as, in their judgment, they thought they would be able to do anything. The two started early in June; 1 think the 6th. Mr. Chester started first with his boat, but he did not get more than a mile above Cape Lupton, when he lost his boat. He then returned, and as he returned, Captain Tyson started off with his boat. Mr. Chester prepared the Hagleman canvas boat, and started off with that. After they had all gone, Captain Buddington set the rest of us at work to try to get the vessel out. We sawed the ice for one or two days there, and at last got the vessel so that she would float. As soon as she floated she began to leak much worse than she did before, and we were compelled to keep the little engine going. We took a trip out after we got the vessel adrift to try to catch up with the party that had gone off in boats. We thought they might be half-way to the north pole by that time; not having heard anything from them, we thought we would try to reach them. We coasted along and got very nearly up as far as Cape Brevoort, but we found the ice packed close and heavy. We several times sailed up and down along the edge of that pack all the way from the east to the west coast. Finding no chance of entering, we returned.

A few days after the return from our voyage, two men came over from Mr. Chester's boat, and they told us that Mr. Chester had gotten out of provisions and wanted more. They informed us that the boats were up in the mouth of Newman's Bay; that they had not been able to get any farther, and that there was apparently no prospect of getting any farther. Captain Buddington thought if he could get both the crews back to the vessel, we would be able to work the deck-pumps and keep the vessel dry. He thought we would be able to watch the opportunity to get north from the vessel just as well as from the ice-floe up in Newman's Bay, and have just as good chance to get north with the vessel as with the boats. So he kept those two men on board, and sent the native Hans over with a letter to Mr. Chester, telling him to return.

After a while the native returned with Dr. Bessels, bringing a letter from Mr. Chester in which he stated that they would return as soon as they could get their boats down. He said they would wait there until the ice opened, and then they would bring their boats down to the vessel so as to save them. But Mr. Chester wrote again and said that he wanted provisions, and that he would like to have at least one of his men back.

Captain Buddington went out again in the vessel for a third time, I think, and not finding any chance to get the vessel in at Newman's Bay, he landed these two men up as near to Newman's Bay as he could get the vessel, and then gave them some provisions to take over to Mr. Chester. He then told them to tell both Mr. Chester and Mr. Tyson to come over as soon as possible. After a while Mr. Tyson and his crew came over, having left his boat over at Newman's Bay, and shortly after that Mr. Chester and his crew came over. Mr. Tyson came first. After they were all on board we got the deck-pumps started and kept the vessel clear. We kept her clear a good deal easier than we thought we would be able to do; and by dividing the whole crew into three watches we were able to keep the vessel clear by pumping five or ten minutes only in an hour. We did not do very much after that until we started for home. That was the 12th of August, and the reason given for going home at that time was that the vessel was leaking, and we did not have more than enough coal to last us through the winter. If we had staid there our fuel for cooking and warming purposes, and keeping the vessel clear of water, would exhaust all the coal during that winter. Then the next fall, if we tried to get home we would have to trust to sail, and it was not thought right to trust only to sail, as we might not then be able to get down; so it was concluded to start for home that fall. On the 12th of August, although the ice looked pretty close around, still from the top of the hill we could see the leads of open water running down to Kennedy Channel; and so we started. We got down to the mouth of Kennedy Channel, I think, in one day. Then we were delayed a little. I cannot recall every little occurrence, but I remember that we tied up two or three times to a floe, and then started again. We were permanently tied up on the 19th of August. We started down on the west side of the channel, but our leads led us all the time in toward the east until we got to where the leads ended, and then we got stuck. They kept running all the time toward the east. All the leads just happened to be in that direction, and we took that direction thinking we could work out again. We were not able, however, to do it. After we got farther down, I believe on the 26th of August, we made another attempt to get out, but the vessel was not heavy enough to move the floes around. I think that that was the only time where we would have been benefited if we had had a Scotch whaler in the place of the Polaris. They are heavier vessels, and more powerful, and can move larger pieces of ice. They might possibly have gotten out at that time, but we could not.

We drifted very slowly down through Smith's Sound, tied up to a floe. It was quite a large floe, and going down we built a house on the floe, having found a pond of fresh water. We dug holes in the ice and stuck small poles in, and covered them with the old awning that was on the vessel the winter before. We put 15 cwt. of bread in the house. That was to provide against any accident occurring on the vessel. We were just quietly drifting down. Sometimes we would drift a very short distance in a day. As we got farther down we drifted quite rapidly, so that on the 15th of October, the last time we saw the land, we were a little below Gale Point, on the west side of the strait, and were a little closer to Gale Point than to the opposite point on the east coast. That was the last point of land we saw. That will give a little idea of where the vessel was when the ice-party broke adrift. I am not certain whether it was on the 15th or late on the 14th that we saw Gale Point, but I know that was the last chance we had to place the vessel before the ice-party broke from us.

At 6 o'clock on the 15th of October, one of the seamen came running into the cabin and told the captain that the ice was breaking alongside of the vessel. The vessel was fast to the floe on its port side. We went out, and in a short time the ice on the starboard side of the vessel all swept past, and there was open water there. Then, shortly afterward there was ice there again. Whether the ice came against us, or we swept against the ice, I could not tell; but the ice gave us a pretty good squeeze when it came there, especially around the stern of the vessel. It cracked the timbers a good deal, and tilted her up, and there was some considerable chance of the vessel being broken. So Captain Buddington ordered things to be thrown out on the floe. We threw everything out. We in the first place took our records out; that is, Mr. Meyer and myself did. Then we threw over everything that was on the deck. We had provisions, &c., piled on the deck for this special purpose, and we threw them out on the ice. The pressure was so great that it was breaking off pieces of ice alongside, and caused a space there to be filled with broken pieces. As we threw the things out there was danger of their falling through the ice, so a party was sent out to take these provisions away from alongside the vessel, and carry them back on the floe, where they would be safer. The Esquimaux had gone out before, and several of the seamen went out. No one was selected especially to go out, only there was a general call for some men to go out there and help to move these things, and these men went out. I think a little after nine we had thrown everything out, but these men on the ice had not removed everything, because they had not gone out as soon as we commenced to throw over.

There were two hawsers fastened to the stern of the vessel, and one to the bow, and during the first part of the gale one of these stern hawsers was fastened to a cleat on the side of the vessel. The pressure was so great that it just snapped this cleat right off, and then both hawsers were brought to the mainmast and fastened around it. Then toward the close, about 9 o'clock, after a good many things were on the ice, after the boats were on the ice, and the men were on the ice, the floe that they were on began to break up; that is, the edge of it. We supposed that the floe must have been broken just where our stern anchors were in, and consequently the stern anchors drew, and that swung the vessel's stern around and brought all the pressure on the forward hawser. Then the forward hawser seems to have slipped off. As near as I can understand, this piece had been fastened on there a little carelessly; at any rate it slipped off, and that let the vessel get away.

We soon lost sight of the party on the ice in the distance.

Question. Was this a dark night?

Answer. Yes. The moon was trying to shine, but it was not doing much. There was drifting snow and heavy clouds, so that we soon lost sight of them.

Question. This parting was wholly accidental then?

Answer. Yes; that is, as far as our party was concerned. It was entirely accidental unless some person maliciously cut the rope, which I have no idea was the case. We thought from the pressure being so much on the stern that the rudder was broken as well as the propeller. We did not know exactly what to do, but the two men were working the pump in the alley-way—I forgot to tell you why working the pump in the alley-way was necessary. The vessel had been pumped out by a very small steam-pump, for which steam was made in the little boiler, that only required as much coal as a common stove. After this little pump had gotten out of order, the engineers were repairing it, and while the work of repairing was going on, the pump in the alley-way was kept continually going by four men, who relieved each other. After we had drifted away from our companions, they told the captain they did not think they were pumping the vessel clear, because it did not suck; it used to suck occasionally. The fireman went down and examined, and found the water gaining very rapidly. The captain ordered the fireman to get up steam so as to work the larger steam-pump, and then he started the deck-pumps. We worked the deck-pumps for about an hour, but still the water was gaining a little on us; only a little, however. In that hour the firemen were able to get up steam. Just as we got up steam the water was running over the fire-room floor; but as soon as we succeeded in getting steam up, the steam-pump kept the vessel clear. That was then about 12 o'clock. The weather had moderated a great deal; the wind had died away, and the moon came out a little brighter, so that we could see better. We could not do anything, and so we sat up in the room waiting for daylight. We could not get to sleep. We had thrown all our bedding away out of the cabin. When daylight came on the 16th, we found that we were in young ice about four or five miles from the shore, and on the east coast about two miles above Littleton's Island. As soon it got to be daylight, so that we could see pretty well, Captain Buddington sent Mr. Chester up to the masthead with a glass to have him ascertain if he could see any of our comrades who had floated away from us. He came down and reported that he could see a piece of ice astern of us out in the straits; that there was something that looked like barrels, or boxes of provisions, but he could not see any signs of the men. That satisfied us all. The reason it satisfied us was because we had an idea that the wind drifted us away from them, and that the current acted against the wind, and took them down, or at any rate did not permit them to follow us, the consequence of which was that there was a great distance between us. We had no idea at all that any one could see them. So when Mr. Chester came down and reported that he could not see them, it just satisfied us at once that they were too far off to be seen. That is the reason no one else went up to look.

A breeze sprang up pretty soon, which broke up this young ice and made lanes through which we worked the vessel into the shore; and we ran her aground. As soon as she was aground, at low water, we looked at the bow. The lower part of the bow was broken off entirely. It was just lying alongside the port side of the vessel. It was still fast, but just bent around. We had thrown away a great deal of coal, but we had four or five tons in the bunkers. Of course we could not, with only these four or five tons, keep the vessel afloat, and so we concluded that the best thing that we could do was to build a house on shore. So we went to work doing so. The natives came and helped us, and in four or five days we were all on shore. The vessel was then abandoned. Everything valuable that was in the vessel, before she was allowed to fill up with water, was either taken ashore or placed on the upper deck, so that we could return and get it if we needed it. Everything that could be used, in fact almost all movables, of whatever character they were, were taken off. We then lived in this house until the 3d day of June. Then we started down in the boats that Mr. Chester had built with the assistance of Booth and the carpenter, (Mr. Coffin,) out of the Polaris.

We started on the 3d of June. On the 23d of June we had gone as far as about twenty-five miles southeast of Cape York, and there we saw the Ravenscraig, and abandoned our boats, and took our personal effects and walked over the ice to her, a distance of about eight or ten miles. We were received by the Ravenscraig people very kindly. That ended the expedition.

Question. Was there any other attempt made to ascertain if you could see the men on the ice except Mr. Chester's going up to the masthead that day?

Answer. No, sir; that was all except what a person could see from the deck. No one saw them from the deck. There were several looking from the deck for them.

Question. Didn't anybody else go up to the mast-head?

Answer. No, sir; not that I recollect of, and I think I recollect pretty accurately about that, because I remember I reproached myself all winter because I did not do it.

Question. Didn't Henry Hobby go up?

Answer. No, sir; not that I know of. He might have gone, though, but I do not remember that he did. We were pretty busy there during the remainder of the day. He might have gone up when I was not looking, but I do not remember of anybody else but Mr. Chester.

Question. Would you have been put to any inconvenience if you had picked up the lost portion of the crew with the provisions and stores? Suppose, for instance, the stores and provisions had been lost on the ice, and you had taken back the men; would it have been at all difficult to have supplied them with provisions?

Answer. In the first place we would have had to practice a little economy, and in the second place we could not have been so generous with the natives. Otherwise I think we had the means to provide for them.

Question. The men on the ice had as much stores as you had, had not they?

Answer. I do not know. I could not tell. They had a great deal.

Question. Did you have any interesting personal adventures after you were put on board the whaler to come home? Why were you delayed so much longer than those who were on the Arctic in getting back?

Answer. I could not tell exactly. The Ravenscraig divided the crew. There were about seven on the Arctic. Then the next vessel they met was the Intrepid, and the captain came on board, and they got to talking, and then decided to put three on board the Intrepid. The captain of the Ravenscraig signified which three he wanted to go on board the Intrepid. Before that he had tried to make us decide for ourselves who should go. We went on board the Intrepid, Mauch, Booth, and myself. We remained there, I believe, until the last of August. Finally the Arctic got ready to go home. The Arctic claims that she ran the ensign up signaling her intention to depart, so as to give the rest of the fleet a chance to send their letters home by her, but the captain of the ship we were on declares that he did not see it, and that he was watching him every once in a while during the day, because he thought that the Arctic was pretty nearly full, and that she would be going home soon, as the captain is a man who always does get home if he possibly can before the others. The captain's name was Adams. The captain of our boat did not see him put up the flag, but he saw him go up by the Ravenscraig, and saw boats passing between them. It was some distance off, but then he had a very good glass. He could not, however, state positively whether our men had gone on board or not. That was the last we heard of it, until we met some parties that had been near by

11 P

or that had been on board the Ravenscraig or the Arctic, and found out about it. They told us that the Arctic had taken the Ravenscraig's people home.

Question. Did the Intrepid make a course in and about the same direction as the Arctic? Did you go over toward Parry's encampment, or did you go in a different direction?

Answer. No; we went in Prince Regent's Inlet, down on the regular whaling ground there. We did not go quite as far as the Arctic. The Arctic went up past Fury Beach. They landed there and got Parry's provisions, &c. We went nearly up to Fury Beach, but did not see any particular use in going there, because there were no whales there. We then came back.

Question. Did the Intrepid get a good supply of whales?

Answer. Yes; she got a fair supply. She had nineteen whales when we went on board, and she got five afterward, which made twenty-four; about 163 tuns of oil, while she could only hold 170. So she did very well indeed.

Question. So she did not lose anything on your account?

Answer. No, not at all. Nobody did. The Ravenscraig did not lose anything on our account. She was, however, unsuccessful, procuring only one whale, and that a dead one, producing only three or four tons of oil.

Question. Did the captain go out of his way on your account?

Answer. No, sir; not a particle. He did not have to leave, either, on our account. He could have staid there as long as he pleased as far as we were concerned.

Question. Did the Eric have any of your party on board?

Answer. We staid on the Intepird until the 24th of August, and then the Eric got ready to go home. She ran up an ensign. As soon as that was done the Intrepid bore right down to her—not for the purpose of sending us home, but for the purpose of having letters taken. But after they had gotten the letters, then Captain Walker, of the Eric, said, " I am going home, and if these Polaris men would like to go with me, I will take them." Of course we wanted to get home as soon as possible, and we went with them.

Question. How much later did the Intrepid get in than the Eric?

Answer. About two days.

Question. How much later did she start?

Answer. She started a good while later. She passed Cape Farewell four days before we did. The reason of that was this: If you will remember, there was a company sent out to find some minerals out in that section, and the Eric was chartered for the purpose. There is still coal there, 85 or 90 tons, and the Eric generally puts in there for fuel. So that instead of going right across Davis's Straits to run for Cape Farewell, she ran down the coast in order to make Exeter Sound, but when she came opposite Exeter, she found that there was so much ice that she could not do it. Then the southwest or southerly breezes commenced and kept her up there, so that she could not get out. She did not want to use steam. Whalers generally try to get home under sail. In the first place the coal is pretty well used up, and then they need some to work around the coast of Scotland. So she could not get out of there for a long time, and the Intrepid got around Cape Farewell four days before she did, but the Eric got home a day or two ahead. That, however, is the reason of our detention.

Question. You were treated well on board these ships?

Answer. Yes, very well indeed.

Question. Was anything supplied to you by anybody?

Answer. The captain of the Intrepid gave me a little clothing out of his chest, and then I got a pair of shoes from one of the men. The captain said he would have that settled. I told him to send the bill in to the American consulate at Dundee. I have not heard anything from it, but I expect to.

Question. Where did you mess?

Answer. Aft with the captain, mate, and the doctor. The other men messed with the ship's company. In the second vessel they messed with the cook. They were given cabin fare on board the Eric—given everything they had in the cabin, only they did not come there to eat it.

Question. They all treated you well, then?

Answer. Yes.

Question. Did you know of any difficulty between the people on board the Polaris at any time?

Answer. Nothing serious. There were difficulties, of course, but only such as I think any crew would have under similar circumstances.

Question. Was there any difficulty among the officers—any difficulty between Captain Hall and any of the officers?

Answer. Nothing serious. They had their little rows once in a while, but I never saw anything that could be said to be at all serious. They were just little differences, that was all.

Question. Did you ever, after Captain Hall's death, hear anybody express himself as glad, or as being relieved by Captain Hall's death, or anything of that kind?

Answer. I could not say that I did. I do not remember anything positive. I heard that some persons said that others had said so, but I do not remember that any one ever said so to me. I have often heard, however, that persons had said so.

Question. Who did you hear had said so?

Answer. I heard that Captain Buddington had said so; that Dr. Bessels had said so. I do not know that there are any others.

Question. What did you hear they said?

Answer. I cannot remember the exact words; it was expressing relief as though they had been under some kind of restraint which was not pleasant, and they were glad it was over.

Question. How was the discipline of the ship while Captain Hall lived?

Answer. It was very effective; that is, if anything was wanted to be done, it was sure to be done. Captain Hall, I think, was a very kind man.

Question. Pretty good disciplinarian?

Answer. I could not say that; I do not think that he had very much executive talent; but I know that while he lived he had order and what would be called discipline—that is, everthing he wanted to have done was done.

Question. Was anything done that he did not order; was anything done that was in any form a violation of the spirit of his wishes?

Answer. O, no, I think not, except temporarily that case of Mr. Meyer's; that was at Disco. He refused temporarily to do what Captain Hall wanted him to do; but that was owing to a misunderstanding all round. It was settled afterward.

Question. How was the discipline after he died?

Answer. Well, it was a good deal easier—that is, there was more freedom, but I think everything that was necessary to be done was done. Captain Buddington was very easy with every one. He tried to get along without having any disturbance or row. I think I can say very safely that everything that was necessary to be done was done.

Question. Was there any difficulty between Captain Buddington and Captain Tyson of any kind?

Answer. No; that is, no difficulty in regard to the business on board the ship.

Question. Did their relations appear to be cordial between each other?

Answer. Yes; after some lengthened conversation, perhaps, there would be a want of some little cordiality, but after a short time they would be very friendly.

Question. Do you know anything about whether Captain Buddington ever got drunk?

Answer. O, yes; he did get drunk, but not very often. I could not tell you how many times he did get drunk, but occasionally he would get so.

Question. Was that before or after Captain Hall's death?

Answer. Both.

Question. This night that you got beset finally in the ice, in the middle of Kennedy Channel, was he drunk or sober?

Answer. I do not know that on the night that we finally got beset that he was, but I know that in coming down there one night he was drunk.

Question. Drunk enough to incapacitate him from duty at all?

Answer. I do not know. It is pretty hard to tell. Some men when they are drunk can do a good deal better than when they are sober.

Question. Was that the case with Captain Buddington?

Answer. I do not know. I cannot tell you. I did not think at the time that he was doing anything out of the way, and I do not know anything that he did out of the way. Of course I believe it would have been a great deal better for him if he had been sober, because I do not approve of people getting drunk. Still I do not think that at any time his getting drunk incapacitated him from doing his work or interfered with the service. There is one thing, however, to be said regarding Captain Buddington, and that is that everybody has been saying he was drunk. It is true enough that he was drunk at times, but it must be taken into consideration that very few glasses will make him drunk, and it is hardly fair, therefore, to talk so much about his being drunk when he really did not drink quite as much as some others did. If he took a couple of glasses they would go right to his head. Of course he did not do right in getting drunk, but I think he is blamed a great deal more for it than he ought to be.

Question. Did anybody else get drunk?

Answer. Yes, there were several that got drunk.

Question. Where did they get the liquor?

Answer. Liquor was on board the vessel, put up under the head of "hospital stores." They took it and drank it.

Question. Did they take it while Captain Hall was alive?

Answer. Yes.

Question. Did they steal it?

Answer. Yes, they stole it.

Question. Did anybody know when they took it?

Answer. In a closet in the cabin there was some liquor that the doctor had stored away there. I frequently saw a person with a key that he had for that go in and get it. Then there were other liquors stored in other places that they got into and got out less openly, because, of course, the place where it was kept was in a more remote part of the ship. When Captain Hall was alive I do not think the officers took much, but I think the men forward got a little, though I never saw any of them drunk at all.

Question. To what extent do you suppose that liquor was used?

Answer. A good deal of it was used toward the last when other material was used up.

Question. After Captain Hall's death was this permitted? It is a question of discipline. Were officers permitted to go and take liquor and get drunk?

Answer. O, no; nothing of that kind. Of course when the officers did go and take the liquor and did get drunk, all that could be done was to accept the fact, and keep them quiet and get them to bed as soon as possible. I do not think that Captain Buddington ever authorized the use of liquor in any way.

Question. If he did not authorize it did he permit it? Did he try to stop it?

Answer. I do not think he made any very strenuous efforts to stop it. I do not know, because I believe that the only way that it could have been stopped by a person who wanted to stop it was by taking all the liquor on board the vessel and throwing it overboard.

Question. Who was it that took the Doctor's liquor in the cabin?

Answer. I have seen Mr. Schumann take it. He was the engineer. He made a key to that door. I do not remember any other one.

Question. Did the Doctor know that he had the key?

Answer. No, sir; not that I know of. I knew it.

Question. Did any of the officers get drunk while Captain Hall was alive so that he knew it after they started from New York?

Answer. I do not know.

Question. Did this habit of taking liquor and getting drunk happen during Captain Hall's life-time, and did he know it?

Answer. No, sir; I do not think the officers, during Captain Hall's life-time, took the liquor. I never saw anything of it, and never heard anything of it.

Question. Then that matter of taking liquor by the officers was after Captain Hall's death?

Answer. Yes, as far as I know, but I know Captain Hall used to miss liquor, because I remember of his opening a box of liquors and finding a bottle or two empty, but it was explained afterward where that went to. It was not known at the time.

Question. Where did that go?

Answer. The men took it. They crept in through the shaft of the engine, and up through there after the liquors.

Question. Then, from your statements, you did not know of any of the officers taking any of the liquor and drinking it while Captain Hall lived?

Answer. No, none of the ship's liquor.

Question. After he died, if anybody wanted liquor, could they go and get it?

Answer. No, if anybody wanted it he could go and get it—that is true enough—but such a one had to watch his chance in order to get it; had to steal it, in other words. The liquor was not at any time put on the table except at Christmas, when we had a little wine on the table, and at New Years, and such festal days as that, they had bottles of wine out on the table, but each one only had about a glass around. There was no time when any one could go and get liquor, unless he stole it.

Question. Where was the liquor kept?

Answer. It was kept in different places. Some of it I say was stored in this closet in the cabin, and a good deal of it down in the hold—down in the lockers in the hold down aft; and I think Dr. Bessels had some in his room.

Question. Alcohol?

Answer. Alcohol and other liquors too, I believe, but I do not know positively about that. I know he had some alcohol, because he showed it to me. His bed was fixed up so that below his bed there was an opening. I know that he showed me some alcohol that he had stowed away in there in his bunk. Then, besides that, there was some liquor in the main hold of the vessel down with the rest of the provisions.

Question. At the time when you separated on the ice was Captain Buddington drunk or sober?

Answer. Sober.

Question. Did he destroy the liquor that was left after you got up beyond Littleton's Island himself?

Answer. Well, I think he did it himself, but I do not know. He might have got somebody else to do it, but there was some alcohol left, and one or two got tight, and then Captain Buddington said there was no use in this thing; that if we had to live there we must have sober men to live with, and so he just went to work and destroyed all the alcohol he could find.

Question. Did the doctor make any remonstrance about that?

Answer. I do not believe he knew it. It was the only way of doing it. We could not expect to have it around there and the men not get at it. I believe the doctor did medicate several of his cans—put in them some tartar emetic.

Question. When you separated on the ice on that night what became of your records?

Answer. I put them out on the ice the very first thing.

Question. What did they consist of?

Answer. Astronomical observations that I had taken up to September 5th of that year; also the magnetic observatious; that is, the observations for the variations of the magnetic needle, which were continued hourly for about four months. Besides which they contained all my dip observatious, and all for absolute determination of the variation; all the observations that I had made for that purpose, and they contained, also, some observations for relative intensity and for absolute intensity, and I think a few for the absolute declination made with the magnetometer. They were all in the box. Among other things there was a collection of plants. I had several specimens of every species of plants that were found there. I think I had one plant that Dr. Bessels did not have in his collection. He had three or four grasses that I did not have; but I had a little plant that he did not have.

Question. What was it?

Answer. I do not know what the name of it was.

Question. Were Dr. Bessels' saved?

Answer. Yes; everything that Dr. Bessels had was saved, except a very few papers. In fact, I do not know that anything of his was lost.

Question. What observations did you take; what was your work?

Answer. Astronomical, magnetic, and pendulum observations.

Question. And the records of all these were lost?

Answer. All except the pendulum observations, and those Dr. Bessels took charge of.

Question. Have you any means of reproducing any of the more important results or data of this series of observations? Have they gone out of your mind? Have you made any reductions?

Answer. I have not been on very many different points, that is, of any particular importance, except when at our winter-quarters, and the two points that I visited on the sledge-journey. The rest of my observations merely gave the position of the ship as she was drifting down. Then the magnetic observations—the absolute variation—I have got that; that is, approximately. I worked up a few of the observations.

Question. Shall you be able to work out any facts in regard to these matters that are not provided for in Dr. Bessels' labors? Have you any data? For instance, have you the means of securing any data in regard to the physical condition of the North that Dr. Bessels has not got in his papers, or that you have not already communicated to Dr. Bessels, and of which he has made use, or can make use?

Answer. When Dr. Bessels and I went down on that sledge-journey to the south, after we separated from our sled we walked, and I carried the theodolite. Then the observations that I made at that time—that is, the bearings of the different places for the determination of the positions—were placed down in Dr. Bessels' journal by himself. I took them, and he worked with me and placed them down. That little work—it did not amount to much—it was only three or four points that it was intended to establish—but all these bearings that I took depended for their value upon the height of a mountain near by. The height of that mountain Dr. Bessels has not got.

Question. Have you?

Answer. It is only from my memory; that is all.

Question. Have you any data that will tend to perfect or improve the map that the Hydrographic Office has made from Dr. Bessels' observations?

Answer. No; I have given them all the information I have on that subject. It is merely from memory.

Question. What is the character of the property in the apparatus and material that was taken ashore from the Polaris to your winter-quarters on Littleton Island? Have you made out at all a list of the instruments that were saved? Were all the instruments saved and removed from the vessel?

Answer. No; not all the instruments. The transit instruments and the pendulum instruments were saved, beside the chronometers.

Question. And what was left on board?

Answer. No instruments whatever.

Question. Or records?

Answer. None.

Question. You removed everything?

Answer. Yes.

Question. Did you bring off with you, when you embarked the following spring, all these things, or did you leave some behind?

Answer. No; they were left behind.

Question. Did the Tigress bring back some of these things?

Answer. No; they did not find these instruments.

Question. What became of them?

Answer. They were back a little distance inland; perhaps a quarter of a mile from the house.

Question. Didn't you tell the Esquimaux?

Answer. Yes; the Esquimaux were right around, saw us put them there, but then the Tigress had no means of finding out where they were except by interviewing the natives, and they did not stay long enough for that purpose. I think I can answer there is not much there. We did not leave anything valuable. The log that has been spoken of, that was preserved, was the log that Mr. Chester wrote. He found he had made a mistake in the first one; that he had left out a day in it, or something of that kind, and so, instead of correcting the mistake, he started a new one and copied the whole thing up to that date, putting in the day he left out. Then he kept on writing the log. He had two large books. Then he condensed these two large books into a log-book, that he brought back here, and these two large books were buried when we left the house, with the instruments. The old log, that had been copied twice, was left knocking around the house.

Question. That was probably brought in?

Answer. I presume that is the one that was brought, because the other two were carefully wrapped up in oil-silk and put in the trunk with these other books.

Question. It never occurred to you, I suppose, that a vessel might be sent up to your winter-quarters?

Answer. As our time of absence had not expired we had no reason to suppose that the Government would send a vessel to our aid. We thought they might do such a thing, but concluded the best thing for us to do was to look out for ourselves. As soon as we found, by going on board the whalers, that the other party was picked up, we were then certain that a vessel would be sent out.

Question. Would it have been better for you to have remained until the Tigress came up?

Answer. I do not know whether we could have remained there that long. It would have taken pretty strong discipline to have kept some of the men there. If there had been a certainty of a vessel coming, possibly it might have been done. You see it would be two months thrown away, because during those two months we might have reached the settlements, whereas, by staying there two months, and no vessel coming, we might not have reached the settlements.

Question. Were not you exposed to considerable danger by sea?

Answer. We had not been, up to the time we were picked up. It is true, the worst part of our journey for our boats remained to be gone through with, because, as you come down farther south, you meet with less ice, and consequently have heavier sea, whenever the wind blows.

Question. Do you know what became of Captain Hall's journal?

Answer. After his death all his documents were put in a large tin box, and kept there, and that box was put on the ice with the other things.

Question. You remember that?

Answer. I did not see it actually put out, but the man who had charge of it (Mauch) told me he put it out, and I believe him.

Question. After Captain Hall's death, was there a formal examination of his effects?

Answer. There was not.

Question. No sealing up or taking of an inventory, or anything of that kind?

Answer. None that I know of. I remember the day Captain Budding-ton looked over his things. He looked in all the different places and around the different desks, and put everything in this one box. He did it there in the cabin. There was no formality about it.

Question. Have you any idea that Captain Hall died from anything but natural causes?

Answer. No, sir. I have no reason for believing otherwise. I believed at the time he died from natural causes, and I have had no reason to change my mind since.

Question. Have you seen this chart that Meyer made?

Answer. Yes.

Question. What general criticism have you to make in regard to that?

Answer. I think he has got Cape Constitution a little too low down. I think, also, that he has got too great space between Franklin Island and the island four or five miles off the west coast of Kennedy Channel. Then the coast above Cape Constitution, and between that point and the southern shore of the southern fiord, is not very accurate, according to my recollection. The reason of that is that Mr. Meyer was never there. This was traversed by Dr. Bessels and myself.

Just on the promontory represented on Meyer's map between Kennedy Channel and the southern fiord there is another small fiord, which runs in for about twelve miles. I did not go into the fiord. Dr. Bessels went up to the top of the fiord. He told me it was about twelve miles. And there is another small island besides the one represented on the map near the shore at that place. It was on that island I took the observation, and made the latitude 81° 05′ north. I never was on the land above our winter quarters farther than a day's walk—just a little above Cape Lupton on this high land.

I do not know of any other criticism I have to make on the general outline of the map. I was not on the land myself higher north than about Cape Lupton, as I have said, and so my knowledge of the lay of the land above that is only general.

Question. What sort of animal life did you find up there on the land?

Answer. We brought in rabbits, foxes, musk-ox, and there were quite a number of little lemmings running round. We noticed their peculiar track on the snow in the course of the winter. We did not know what it was for a long time, but when the spring came we caught a great many.

Question. Any blue foxes, or all white?

Answer. I think we only got one white fox up there. That is all I have any remembrance of.

Question. Any wolves?

Answer. There were reports of wolves having been seen, but I think it is very doubtful.

Question. Any brown bears?

Answer. None.

Question. Any white bears?

Answer. The doctor found a white bear upon this fiord I was telling you about, which extended twelve miles in. He found a white bear there that Joe killed. That is the only bear we met with on the expedition.

Question. Any white partridges?

Answer. Yes, sir; quite a good many in the spring.

Question. Did you kill many of them, and eat them?

Answer. I never had but one mess. I do not know how many there were. They had a good many messes, I believe, when the doctor and I were gone on the sled journey.

Question. Any white owls?

Answer. There were none caught. I believe Joe saw one.

Question. Any hawks?

Answer. I do not know exactly. I never saw a hawk that I am aware of.

Question. Any eagles?

Answer. None.

Question. Any sea-gulls?

Answer. Yes, a great number of gulls, of different kinds. We used to see the burgomaster. That is the large gull. Then we saw what, I think, are called the "swallow-tails." That is the English name for them. They were white on the lower part—the belly—with a kind of grayish, bluish tint on the back. I do not think the top of the head was black. I think both the feet and bill were red. They may have been the Arctic terns, but they were called "swallow-tails."

Question. Did you see any sabine gulls?

Answer. Yes, but they were not common. I killed one, and Mr. Mauch killed another.

Question. Did you see a small gull about the size of a sabine gull—white gill, with a black ring around his neck, and with a wedge-shaped tail—no central tail feather?

Answer. No; we saw quite a number of the gull-chaser.

Question. Did you see any whales?

Answer. I did not see any whales until after we got into Lancaster Sound—not what they called whales. Some people call narwhals whales. I think we saw narwhals first around Whales Sound.

Question. How far north did you see them?

Answer. I do not think we saw them higher than 77°, perhaps 77½°.

Question. Did you see any walruses?

Answer. Yes.

Question. How high did you see them?

Answer. Where we wintered the last time—about 78½°. In regard to those narwhals, when we drifted down in the ice through Smith's Sound, in a hole of water, we saw two fish come up. Captain Buddington said they were narwhals. I remember they went out after them to shoot them, but they disappeared before they could reach them. Mr. Chester said they were narwhals. If they were narwhals, then we saw narwhals a good way above 77½°. That was about 79°.

Question. Were the walruses very abundant?

Answer. The natives seemed to catch quite a number; and in the spring we could see quite a number on the ice.

Question. Did they catch both sexes, or only males?

Answer. They caught females, because they one day brought a little embryo walrus for us to eat. They said that they had gotten it from the inside of a walrus.

Question. Was it good?

Answer. We did not have courage enough to cook it. It was almost too young.

Question. You are sure that you saw males, also?

Answer. I do not think I would know the difference.

Question. You found no fish of any kind up in Kennedy Channel?

Answer. No; we saw seals all the way up. I saw the salmon on board the whalers. That was pretty well down. We did not see any fish of any kind in Polaris Bay. We tried several times to catch fish by throwing lines overboard, but we did not succeed. The sea was full of shrimps, and there were some medusa jelly-fish, but they were not so very numerous.

Question. Any clams, or shell-fish?

Answer. No; we did not see any.

Question. Do the seals live on the shrimps?

Answer. I could not tell you that.

Question. Were there birds in sight during the winter?

Answer. No.

Question. During neither winter?

Answer. The last winter we saw them very late, and very early. I do not think we saw them exactly in the middle of the winter, but we saw the ravens very late and very early last winter, and I suppose they stayed there all the time : but whether they did or not, we have no authority for saying. I will state that very early, when the natives went out to hunt the seals, they found these dovekies. They told us it was their custom to leave their young up there one year, and the old ones would go home. The ones born there would remain one winter, but ever after that they would go home. That is all I know about it.

Question. Is that probably because the old birds change their plumage in the winter, and look like young ones?

Answer. No; the natives could not be deceived in that way.

Question. This (exhibiting a map to Mr. Bryan) is a map of Hayes' Expedition. published by the Smithsonian Institution, January, 1865, in which he has laid down the highest position that he reached, at Cape Lieber, and in going up this sledge journey to Cape Lieber, on the western shore, he states that he saw the open polar sea of Kane, just as it was laid down in Kane's chart, and when he was here at Cape Lieber he saw nothing but open water to the eastward. Now, when you were going up, north of Cape Constitution, and as far north as Cape Lieber, did you see the land on the right-hand, to the east?

Answer. Yes; from the time we entered Kennedy Channel we saw land on both sides of the channel all the way up. When we got to latitude 81° 20' we were in such a position that the east coast shut in on the west coast, so that we thought we were in a bay, but after going a little farther north we opened out Robeson Channel.

Question. You see the position of Cape Constitution, as laid down on this map, is just below latitude 81°, and between longitude 65° and 66°. What land do you place there instead of the Cape Constitution of Kane?

Answer. Well, I could not tell you exactly what land, and I do not know about its being in that particular longitude, but in that latitude there is land there that I do not think is Cape Constitution. I could not tell you what land it is. We never gave any name to it.

Question. When you were on that island in 81° 5', did you see the Cape Constitution of Kane?

Answer. No. When we were on that island, we were right behind a head-land. We had to go around the head-land and come down to another cape, and then we opened out a cape to the southward.

Question. When on that island you could not see Cape Constitution, but you could when you went to the southward?

Answer. Yes.

Question. What latitude did you reach when you went south on the land?

Answer. I think probably we went seven or eight miles.

Question. What latitude did you reach?

Answer. That would make it a little below the latitude of 81°—a very few miles.

Question. Did you see what you recognized as Kane's Cape Constitution then?

Answer. Yes.

Question. Where was it?

Answer. It was still to the south of us. We supposed it to be about from twenty-five to thirty miles; not directly south, but in a southeasterly direction, I think.

Question. Was the land continuous between you and it?

Answer. Yes.

Question. Was there any open polar sea there, between you and it?

Answer. No. I think not. There was just a bay that went round.

Question. Was Hans with you then?

Answer. Yes.

Question. Did Hans recognize this land to the southward as the place he had been at before?

Answer. At first he did not; but he seemed to afterward.

Question. Is the land that you saw to the southward of you, then, the same land you recognized as Cape Constitution, when you came down in the ship?

Answer. Yes.

Question. Did you see Franklin and Crozier Island off it?

Answer. Yes.

Question. Could you see Cape Lieber from Polaris Bay?

Answer. Yes. We could see it all the time, and even in the winter time, when the moon shone. I will state that near where Hayes laid down Cape L. Von Buch there is a island in the channel, four or five miles from the western shore, which is not laid down in Hayes' map.

Examination of witness being concluded, adjourned to meet on Friday, December 26, 1873, at 11 o'clock a. m., at the same place.

FRIDAY, *December* 26, 1873.

Board met pursuant to adjournment.

Examination of Joseph B. Mauch.

I was born in Germany. I am twenty-four years of age. I sailed with the Polaris from Washington in the capacity of a seaman. I went from Washington to New York, and from New York to New London; from New London to St. Johns, and from St. Johns to Fiskernaes; from Fiskernaes to Holsteinburgh; from Holsteinburgh to Disco; from Disco we sailed to Upernavik, and from Uppernavik to Tessiuisak; from Tessuisak for the north.

During the whole progress of the voyage I kept my own private journal, which Captain Hall directed me to keep, and told me to keep it in my own way, and to put into it everything I thought proper. That journal I have here. It is contained in these two books, which I now produce, written in English in my own hand. It was written, as a general thing, from day to day, but sometimes two or three days were written at one time, when no particular incident happened during the interval. It contains everything that happened, as far as it impressed itself on my mind, and gives my opinions and ideas freely. In it will be found a much more detailed account than I am now able to give from memory.

Question. Can you remember the day when you left Tessuisak?

Answer. I think it was on the 24th of August, 1871, about one o'clock. We passed Cape Alexander on the 27th; sighted that at first at three o'clock in the afternoon. After that we struck the course over across Cape Frazier, and went over along the west coast, and went as far north as the ice permitted us to 82° 26'; afterward corrected to 82° 16' by Mr. Meyers.

Question. What did you do on board the ship?

Answer. I kept the journal of Captain Buddington after Captain Hall's death, and before Captain Hall's death I kept Captain Hall's journal. Besides my other business I acted as captain's clerk. I lived in the forecastle during Captain Hall's lifetime, and afterward until about the time of the return from the boat-journey. Then I went into the cabin, because it was handier. I had to take observations continually. The captain said, " You had better live there." Captain Hall told me several times during his lifetime to go there and sleep there, because he needed me in the cabin. He had me there continually. I had to go there every day. But at last I refused to go, for fear it might make some disturbance among the seamen. They did not like me to leave.

Question. Do you remember when you passed Cape Constitution?

Answer. I do not remember the exact date, but I think it was on the 29th. I remember having seen it when we passed it. Captain Hall was on deck pointing it out.

Question. How did you know it was Cape Constitution?

Answer. Because it was pointed out to me. I recollect that I saw the islands in front of it. According to that I took it for Cape Constitution, and Morton and Hans both stated that it was Cape Constitution.

Question. What was there beyond that? Did the straits still continue up?

Answer. Yes; up to Kennedy Channel.

Question. Did you see any open polar sea above that?

Answer. I cannot say I did. Above Kennedy Channel we came to a bay, which we afterward called " Polaris Bay." We sailed on up through these straits then, and came into the bay. At that time we could not make out it was a bay, because we were unable to see the east coast. It was foggy, if I recollect right, all the way up. I remember that I could not see the east coast at that time. The next day, however, I saw it again, and we had gone farther north by a considerable distance. We went up through this bay, and we found what was called by us " Robeson's Straits." We went up into those straits.

Question. How near through?

Answer. As far as I could see. We had not clear weather at that time, and therefore I cannot say how far.

Question. Why did you not go farther than you did?

Answer. We got beset. They intended to go farther, as far as I heard, and Captain Hall called a council of the officers and it was decided that they would go farther north. Dr. Bessels's counsel was accepted of going over to the west coast, as navigation would be better there, and sledge journeys, too. They tried to go over to the west coast. Captain Buddington did not like to take that lead, because he was afraid that he would get beset, and in the evening we did get beset in the channel. We drifted for three days.

Question. Before you got to the west coast did Captain Hall make a landing?

Answer. Yes, sir; he made a landing on the east coast?

Question. More than once?

Answer. He made a landing on Cape Frazer. I think it was twice he made a landing. I think he went there first on the ice, and the second time with the boat, but I am not certain about it. He went to look for a harbor. He found one to suit him, but when we tried to get there the ice closed. It came into the harbor and we could not get in. He called it "Repulse Harbor." It is above Cape Frazer.

Question. Was this the day before you tried to go across the straits, or the same day?

Answer. I do not remember whether this landing of Captain Hall was the same day we tried to cross the straits, or the day before.

Question. When you got beset in the ice did you put provisions out? Answer. Yes, sir; we did so for fear the ship would be nipped. We drifted down with the current. I think we were in the ice two or three days at that time. I think the dates are in my journal. The first opening occurred at Polaris Bay where the basin widens. The ice gave way and the consequence was we could reach the east coast. We then put into Polaris Bay. We made a harbor there. It was then decided to go into winter quarters at that place. I do not think we were there but a day when Captain Hall said he would go into winter-quarters, because going up I had to work in the fire-room, and as soon as we got there and dropped anchor, Captain Hall called me up and said my work was done there (in the fire-room) for the future. So I think he concluded at that time that he would make winter-quarters at that point. We landed provisions on shore and put up an observatory and went into winter-quarters. After that there were sledge-journeys sent out. There was one sent down south, Dr. Bessels and Mr. Chester. Mr. Chester was in command. Then Captain Hall made a journey himself. I think he left on the 11th or 12th of October. He was absent until the 24th, I believe. He returned at that time and was taken sick. He was gone about two weeks. I saw him when he came back just before he stepped into the cabin. I had some little conversation with him. He asked me how I had got along, and I told him very well. I had to keep a record during his absence and I gave him that, and he said he would look at it to-morrow, or so; that he was tired then. He said that he was tired and it was pretty cold. That was about all he said. I remember of having said to him "You had a hard time of it;" and he said "Yes." Then he went into the cabin, and I was told afterward that he was taken sick. I was not there at the time he was taken sick, but shortly afterward I went in to see him, to see what was the matter with him. I think it was Mr. Morton who told me he was taken sick. I saw him when I went in, but he did not say much. When I went in he was just undressing, and they were putting him to bed. The next morning I heard that he was in a very bad state. I did not say anything to him when I went in. He was very much occupied then. I think Dr. Bessels was there, and Mr. Morton who was undressing him. I do not remember who else.

Question. When you first went in was there anything in particular that you noticed?

Answer. No, sir; nor did I hear anything either.

Question. Did you know anything about his drinking coffee?

Answer. I heard that he had been drinking coffee, but I did not regard it as anything extraordinary. I thought that it was very natural for him to take coffee after he came back.

Question. Had the men been taking coffee then, or about that time; I mean the men on board the ship?

Answer. No, sir; they were not taking coffee at the time. It was specially prepared for Captain Hall, or, rather, for the party that returned.

Question. What appeared to be the matter with him when you saw him, after he was taken sick?

Answer. He seemed to have been vomiting just before. That is all I know. He did not say anything. He got suddenly worse during the night. I heard he was unconscious then. I never went in when he was very bad. I do not recollect the precise date when I saw him after that, whether it was one or two days afterward, but I think it was two days afterward—may be one day. I saw him again, and he was very sensible. He seemed to be very weak. The men always sent me there to ask him how he was getting along every day, and I inquired every morning, and he was always very much gratified by hearing from the men. As long as he was well I remained in the cabin. I had to remain there. And one day he dictated to me about that report that he left at Cape Bree-vort. I copied that dictation afterward. Joe and Hans had shot a seal, and I could not get clearly their story, and he was there and had them both tell me. He asked them to do so. He was very sensible then. That was a few days before he died.

Question. Was he out of his head at all during his sickness?

Answer. Yes, sir; but I never saw him when he was out of his head. It generally came on during the night.

Question. Did he accuse people of trying to kill him?

Answer. I never heard it, but I heard others say so. I was only there when he sent for me to do work, and that was when he was most rational.

Question. Did he accuse almost everybody of wanting to kill or poison him?

Answer. I do not know that. He seemed to accuse those who were around him at the time. For instance, sometimes I recollect Mr. Meyer told me that he would ask him to protect him, and the next morning he would call him a murderer or something of that kind, claiming that he wanted to take his life. I was not present with him when he died. 1 was with him the day before, in the morning at 10 o'clock, and heard his last words—that is, what I thought were his last words, and I think they were. I am not certain. He only said that he was going to get up and go out. Captain Buddington said "you cannot do that; lie down," and he kept him quiet, and kept off all excitement from him. I did not like to see him suffering so. He was suffering very much that day. He was breathing very hard, and was in a half unconscious state. He was in a sort of stupor. I heard that he was so all day, and all that night until he died. He died in the morning about 3 o'clock.

Question. Did you ever have any idea that he died anything but a a natural death?

Answer. I never had any other idea. I did not have any other idea then, and have not now.

Question. Did he at any time, when he was apparently in his right mind, accuse any one of wanting to injure him?

Answer. No. He never did when he was in his right mind. He never said anything about anybody. I was present at the funeral. With regard to the medicine he received from the doctor, he asked me at the time what it was, knowing that I understood something about it. It was sulphite of iron that the doctor gave him one day for his stomach. He asked me what it was, and I told him how it was prepared, and everything about it. He asked me what a dose was, and I think I took twenty drops, so as to assure him that it was not at all dangerous. I took it right in front of him where he could see me do it. He was content, and took it after that. The doctor had ordered him to take only five drops. He asked me what a dose was, and I told him about twenty drops for a grown person; and I took the twenty drops to show him that I was not afraid to do it. After I took the twenty drops he was willing to take five. It has a kind of inky taste.

Question. Did you ever hear anybody express himself as relieved by Captain Hall's death?

Answer. No, sir; I think not. I do not recollect that anybody expressed themselves in that way. Nobody ever did in my hearing. I heard others say that they had heard such remarks, but I heard nothing of the kind.

Question. How was the discipline of the ship during Captain Hall's lifetime?

Answer. He was strict, but very good and kind to everybody. He commanded respect from everybody, and he paid every one his due respect. The discipline was good, so far as I know, during his lifetime. I do not know that there was much difference after his death. They never enforced the strict rules about putting the lights out, and things like that, after his death, but we just lived on as they do on board a whaler. Captain Buddington commanded respect for himself, and everybody paid him the respect that was due him, with the exception of some of the officers who thought probably they were a little higher, and had more right to exercise command.

Question. Did you keep Captain Hall's journal for him?

Answer. I did not keep his. I kept my own, and he always took his from mine; but he never wrote very much in his journal. He only wrote every few days. There was not much writing by him. During his disease he made me copy that which he had written.

Question. You spoke of this record at Cape Brevoort. Had he it copied off?

Answer. Yes; he had it copied off with pencil, and he dictated that to me, and I wrote it. I wrote it in the cabin on a piece of foolscap paper. I have seen the copy of the record in Captain Tyson's examination. It is a true copy. I myself wrote down a copy of it at the time. That was spoiled in some way or other; some ink got spilled on it, and I put it in my pocket, and kept it in my pocket until last winter. I had forgotten I had it in there. It was an old coat that I had put aside, and last winter I took the coat off, because I did not have anything else—any clothing. So I had to take that old coat, and in the pocket of it I found the old copy from Cape Brevoot. I intended to take it back, but I did not do it. I have since intended to bring it down here from New York, but having seen the same thing in Captain Tyson's examination, I thought it hardly worth while. He dictated to me from the original, which was in pencil. I put that dictation among his records, and they got lost. I had them all together, but that other piece of paper found in his writing desk was the original paper from which he dictated to me. He had brought that with him from Cape Brevoort. This dispatch, as published in the original report is correct. I took it before I left New York, and compared it with the copy I had. It is the very same.

Question. After Captain Hall's death what was done with these papers?

Answer. His papers were put in a large tin box. Captain Buddington gave them to me to put in there. I selected every one I could find to put it in. At the time we got separated from the other party I had these papers. I took care of them, because I wanted to save them. They were Captain Hall's papers, and I wanted to preserve them. I therefore put them out on the ice, and told the men to put them up on a high point that was there. I put them out myself. I do not know whether they were put on the high point or not; but I told them to put them there. Everybody was in such a hurry that probably nobody heard me speaking about it. I went out there myself afterwards, but I never got that far out, because I had to carry provisions from the ship to a point a little farther on, then another picked them up and took them still further.

Question. Was there any regular examination of these papers in having them sealed up after Captain Hall's death?

Answer. No, sir, no formal examination; there was no inventory made of them. I do not know whether any of Captain Hall's papers were destroyed or not, but I do not believe that any one has been destroyed, because I never missed any; and I know what the papers were. No part of his journal was destroyed as far as I know. Everything was put away. After Captain Hall's death Captain Buddington took command, and I still acted as his clerk.

Question. What occurred after that of importance?

Answer. We broke adrift from our winter quarters, and went as far as Providence berg in a gale of wind. I do not think the ship was damaged at that time, but it is my opinion that she became damaged on the 27th of November, when that ice-berg moved in. I think she got crushed then. I cannot tell how much she was damaged at that time. Nobody could. We could not see so as to ascertain. She did not leak at that time because the water was all frozen. She got on to this iceberg and laid there all the rest of the winter. We got away in the following summer. I think it was on the 27th or the 29th of June.

Question. Were any more attempts made to go north during the winter while you laid there?

Answer. There was no attempt made because we had open water out in the straits, and we could not go over the land. Captain Hall proved that in his journey, and after his return, he so spoke. He said that we could not go over the land on account of the deep ravines, across it and the rocks. And he said there was no snow on the land in many places. He said, therefore, that he should try to reach the west coast, the following spring; and go up on the sleds, and go up farther. During the winter we had open water out in the straits and no ice, and therefore no attempt was made to go north, because we saw we could only reach it in boats. That is what we supposed, that we could only get north in boats. We had open water, even in April, and I think in May.

Question. In the middle of the straits?

Answer. Yes, sir; and up at Cape Lupton. It was close on to shore, and we could not get around the third cape north of us on account of the open water. There was no icefoot, and we could not travel over the ice on foot. In the spring two boat expeditions were sent to the north and they were gone some six weeks, and while they were gone we started with the Polaris three times to get north. We supposed the boats to be about Cape Union—what we called Cape Union. It is a little farther than Hayes' Cape Union from where the land turns north on the west coast. We found, however, that we could not get north with the Polaris. The ice was stretching across from Cape Sumner up to the west coast. The result was we never got north. Then those who started on the boat journeys returned without the boats. They could not get back with the boats. We then left Polaris Bay on our way south. I forgot to mention the expedition that was made down south during the winter on the sledges. Dr. Bessels and Mr. Bryan and the two Esquimaux composed the party that started on the expedition.

Question. When you got out with the Polaris to go south how was the ice then?

Answer. As far as I could see toward the north there was all ice, and when we got out there was so much ice outside of Polaris Bay that we thought she would not get through toward the south. We tried at several points to get through. We looked toward the north and we saw nothing but ice. We were satisfied the ship could not get north then. We had not much coal, and Captain Buddington did not think it advisable to go any farther. This was in August. The ice was not as open as it had been the previous year. We tried to make our way to the

12 P

southward then and got jammed in the ice. We first got beset opposite that island on the west coast. I do not recollect the name, but not far from Cape Constitution. That island has no name, I think. It is about in the middle of the straits, if I recollect aright, nearly opposite Cape Constitution. It is a little farther north. We drifted I think for about a day and night. During the night Captain Buddington tried to get through the ice. He started fires again and tried to get through, but did not succeed. The next day we got out into open water. There was a little open water to the south and we were able to steam ahead until we got fast into the ice again at the north end of Smith's Sound, near the south end of Kennedy's Channel. We remained fast, drifting south. We made several attempts to get out, but we could not go anywhere. The ice was closing in on us and we could not break out.

Question. What took place?

Answer. We drifted down a little south of Cape Alexander, and got a gale there. The ice broke off that night in a heavy gale from the south, and the vessel got jammed and got nipped, and we pitched everything on to the ice. That was the 15th of October. We had built a house on the ice before this.

Question. You were still fast to the same floe that you made fast to up in Smith's Straits?

Answer. I do not know whether it was the very same or not, because we changed our position several times. I do not think it was the very same one that we made fast to first. We tried several times to get out, and succeeded in going to the northwest, probably a mile or two, and then we had to make fast again, and could not get any farther. We made attempts to get out of the pack, but could not do it.

Question. What caused the ship to break adrift that night?

Answer. It was the gale that drove the ship off. The anchors came out. That is, the floe broke right where the anchors were planted in it.

Question. Were you all on deck when it broke adrift?

Answer. I was not on deck; I was in the cabin fixing up my box. I had two boxes of clothes that I had just put out, or rather gave Mr. Bryan to put out. When he came in I was just wrapping up some books—Captain Buddington's journal and my own. That is all I had.

Question. Did you hear Captain Buddington order the provisions and stores to be landed on the ice at that time?

Answer. He ordered the provisions to be landed. I think it was at seven or eight o'clock. It was pretty early in the evening. We did not break off until nearly ten o'clock. We had the boats out already, and were helping over things. I was out there when they put the boats farther out on the ice. Then I went aboard, and the next minute we drifted off. I went on board to get the books. I had everything else on the ice.

Question. Then the separation was accidental?

Answer. Yes; entirely so. A number of people were left on the ice, nineteen, I think, and all the boats and sleds. The ship drifted to the north; that is, she drifted until the next morning. When it became clear we found ourselves in a bay—in what we supposed to be a bay. It is above Littleton Island. During the night we had to pump the ship. Soon after the separation Mr. Schumann came up, and said that if he could not keep her free now, she would have to go down, because the water was above the platform and already in the fire-room, and nearly extinguishing the fire. He succeeded in keeping her free with the steam-pump. In the morning, when we found ourselves in this bay, we sailed in. We only used steam to round a certain point of ice.

We did not have coal enough to start steam; that is, to get enough steam for the propeller; only enough to pump her, and just to give a few revolutions of the propeller, so that we could round the point of ice. That we could not do with the sails alone.

Question. Did you see any of the ice-floe that you had been fast to ?

Answer. No, sir. I looked for it from the deck, but could not see it. I looked all that morning when we went in, and that forenoon, and looked about noon, but saw nothing at all. I was told by several that there was a black point there that they supposed was the provisions, but that they could not see any men. They gave me a glass, but, although I have as good eyes as any one, I could not make out anything. I did not go to the mast-head. Those people that told me about the provisions were always seeing something when they wanted to see it. Mr. Chester went up to the mast-head, and was up there continually as far as I remember, with a glass. He saw nothing of them. He thought he saw some provisions, but nobody with them. Then we ran the ship ashore. The stem was broken off below the six-foot mark. I could not see anything of the lower part of the stem.

Question. Was her bow open then ?

Answer. I never saw that the bow was open; I never saw it, because the ice was all around it, and frozen,

Question. Where did you suppose all the water came into her ?

Answer. Forward, at the bows. I know the water got in there from hearing it down in the fore-peak.

Question. Have you been brought up as a seaman ?

Answer. No; I have been a druggist. I passed my examination in New York, in the College of Pharmacy. I have not had a seaman's experience.

Question. Is it your opinion that it was necessary to run the ship ashore to keep her from sinking ?

Answer. She was run ashore and remained ashore, and we left her there. We left her a few days afterward, and built a house. We felt that it was necessary to abandon the ship; that we would never get her out again. I thought so when we came down. In fact we knew that we could not get out of the ice and snow, that we did not have coal enough to steam; and I have been told that we could not proceed with the sails, and I do not think myself we could have from what I saw. We built a house and remained there during the winter. The Esquimaux came to us and assisted us. We were on friendly relations with them. They remained there, off and on, during winter. They built snow-houses and snow-huts around our house, and remained there in that way. Sometimes they went down to the settlement and remained there. Men and women and children, all together, were with us. And I think we had as many as 101 at different times.

Question. Were they there when you left the ship?

Answer. No; they came the next day. We thought it was the party whom we had left on the ice.

Question. When you left the ship in the following spring were they still there ?

Answer. There were some—a few. There was one family that lived with us all winter—that was there—and another family that had come back fourteen days before that.

Question. You lived comfortably during the winter in the house ?

Answer. I could not say we lived very comfortably, but we could not expect anything else under the circumstances.

Question. Comparatively comfortably ?

Answer. Yes; in the spring we began to build boats, and in those boats we left for the South. It was the 2d of June that we left. We all left at the same time, taking with us what we could carry. We left behind us some potatoes, pork, and dried apples, and some bread. We did not have very much bread. We did not bury those provisions, but left them outside. We left in a hurry. Those provisions were canned.

Question. What did you do with the ship's log-books, and other papers?

Answer. Some of the log-books were buried. The old log-book was buried. Mr. Chester wrote a new one. He copied the old one into a smaller one, that it would be handier for carrying, and he took that along. The old log-book was buried, and the instruments also. Some of Captain Hall's books were also.

Question. How was the place marked where they were buried?

Answer. I have not been up there myself, and I do not know. After we started we were in the boats from the 2d of June until the 23d of June—about three weeks. We sometimes slept in the boats and sometimes on the rocks. It was very cold. The first time we had heavy snow. I recollect that when we were on Hakluyt Island we were covered with snow. We managed to keep warm by getting under the blankets and gathering moss together. We had a covering over the boats, but we did not sleep in the boats at that time. I do not know why, in fact, but the covering was not made up until we were on Hakluyt Island. We slept on the rocks, not on the ice. We kept close in to the the Island. Hakluyt Island is close to Northumberland Island. I should think it was about four miles from it. We went on until we got around Cape York, and were picked up by the Ravenscraig, by whom we were kindly taken care of. We were then transferred to the Intrepid and then to the Eric. We then reached Dundee, and from there came here. Mr. Bryan and Mr. Booth were with me. We all came together with the exception of Mr. Bryan, who remained there a week. He had a leave of absence.

Question. In your journal you say that you saw Northumberland Island before you broke loose?

Answer. Yes; I saw it on the 14th, when it was clear. We were a good deal south of Cape Alexander; I recollect that very well. The last point we saw on the west coast was Gale Point. That is the point we were nearly opposite then.

Question. How long was it after you saw Gale Point before you drifted down and saw Northumberland?

Answer. We saw Northumberland Island at the same time, when we were opposite Gale point.

Question. How high is Northumberland Island?

Answer. I think it is about 1,500 to 2,000 feet high.

Question. You had got to the south of Cape Alexander, and had seen Northumberland Island?

Answer. Yes, outside of Cape Alexander. There is one thing I want to say about the observations that were taken up there the first year by Mr. Myers. They were brought back with the records by the doctor, if I recollect right. I do not know whether you have the corrections for the barometers—the barometers that were used. I have some of the observations taken, but have not corrected them. I think you can get the corrections out of my observations. I can send them down here. I did not bring them along, because I did not think it was necessary until lately.

Question. What papers have you now that you kept of the expedition?

Answer. Nothing else I now know of except these two journals, and those observations—the barometrical and thermometrical.

(Mr. Mauch was desired to produce the observations referred to, and he promised to do so on his return to New York.)

Question. Do you remember the day when the Polaris was farthest north?

Answer. Yes. I was on deck.

Question. What did you see when she was the highest in latitude?

Answer. I saw she could not get any farther. There was ice ahead.

Question. What did you see on your right hand?

Answer. I did not see anything on my right hand, because it was foggy to the eastward. I think I saw land to the left. That was when she was up at 82° 16'. Occasionally we have seen land on the east coast when she was at the highest northern point.

Question. How high north did you go yourself?

Answer. I went as far as Cape Sumner—not quite.

(Mr. Mauch produces chronometer of the Polaris, numbered "1381," and having on it the names of "T. S. & J. D. Negus, New York.")

Question. This chronometer went out in the Polaris?

Answer. Yes.

Question. Has it been going ever since?

Answer. Yes.

Question. Had it stopped at all?

Answer. Yes, it stopped at one time in the boat journey. I think it did. It was not in our boat. The doctor had it.

Question. You mean in the last journey?

Answer. I have heard it stopped then. I am not certain about it.

Question. What became of the other chronometers?

Answer. They were left up there. That is the only one brought away, I think.

Question. Have you wound this up since yourself?

Answer. Yes. Mr. Bryan set it again, off Cape Kater, on the west coast.

Examination of John W. Booth.

I was born in Lancaster, England, in 1848. I am a machinist by trade. My first voyage to sea was made in the Polaris. I joined her in Brooklyn. I went in her from there to Greenland, stopping on the way at Saint John's, Fiskernaes, Holsteinburgh, Disco, Upernavik, Kingituk, and Tessuisak.

Question. Did anything of consequence happen on the way there?

Answer. There was only a little accident that happened to the engine. One of the nuts of the reversing link came off. That is all. The blow-off pipe gave out at Saint John's. Otherwise the engine worked well. From the last place I mentioned we went up north without much difficulty. We stopped at Cape Frazer, where Captain Hall went ashore. Then he came back, and we went on without much difficulty until we got stopped by the ice. That, I think, was the 4th of September. We were then in latitude 82° 26'. I was in the fire-room at the time. I was on deck a little while. I saw nothing but ice; no water to the north. I saw plenty of land on both sides, both east and west.

Question. Was that while you were in Robeson's Straits?

Answer. Yes.

Question. Was the engine in good condition then.

Answer. Yes; and we could have gone on had the ice permitted. We had from one hundred and ten to one hundred and twenty tons of coal. In coming along up Kennedy Channel and Smith's Sound, the propeller was making about sixty-five turns. Not being able to get any further north, after some little delay, we went into Polaris Bay for winter quarters, and secured the ship there. On the 10th of October, Captain Hall started off on a sledge journey, and returned on the 24th of the same month. I met him when he came back at the observatory. He said he was glad to see us all again, and he seemed to be in good health. He said he enjoyed his journey very much, though he was a little tired. He then went on board of the ship. From that day I did not see him any more until I went into the room when he was dead. I knew that he was sick.

Question. Did you at the time of his decease have any idea that he died from any other than natural causes?

Answer. No, sir. Even now I believe he died a natural death.

Question. What happened during the winter about your engine and your machinery?

Answer. We took her all down so as to put her together in the spring, and saw that everything was all right, ready for the next spring, to go north. I had that work to do while I was there, and was engaged in making things for Captain Hall's sledge journey.

Question. What did you do during the winter after you took the engine apart?

Answer. I did nothing specially. I occasionally helped the men if there was anything to do.

Question. Did Captain Buddington take command after Captain Hall's death?

Answer. Yes, sir.

Question. Was there any trouble during Captain Hall's lifetime on board the ship, that you know of?

Answer. None that I know of.

Question. Any after his death?

Answer. No, sir.

Question. How was the discipline of the ship while Captain Hall lived?

Answer. Excellent.

Question. How was it after his death?

Answer. Very good. I know of no serious difficulty on board. I had none with any person.

Question. How long did you remain in Polaris Bay?

Answer. We remained in Polaris Bay until the 12th day of August. I forget the day exactly that we went to the north, but it was some time in July, 1872—that is when we made the second attempt to get a passage up. We could not get any farther than the south point of Newman's Bay, on account of the ice. That was while the two parties were away on their boat-journeys. We made a third attempt, but failed, and then we lay in Polaris Bay until the 12th of August, when we turned to the southward. The boat parties had not at that time returned. We had sent for them. One crew arrived just a few days before we started away.

Question. Could you have gone north at that time with the ship?

Answer. No, sir. There was too much ice in Robeson's Straits. Nor could we have gone north at any time after that. When we got into the ice we drifted to the southward, and could not push our way up north. We drifted to latitude 78°, the lowest point, I believe. We were then fast to the ice-floe with hawsers and anchors, while drifting down.

On the 14th of October, in the evening, the storm commenced, and the first thing I knew, while I was working in the fire-room, I heard a crash, and the ship reeled over and almost capsized. I was called on deck, and when I got there was ordered to the pump. We had no steam on at that time, and had had none for three days. The small pump was worked. After being sent to the pump I was called back again and ordered to get the things out of the forecastle, and throw them on the ice, as the ship was in bad condition. Captain Budding-ton ordered me to do so. While doing that he sent me away back to the pump and there I staid until the ship was away from the floe. She had been leaking a good deal for two or three days before. We were at first using a little steam-pump that we had there to keep the water out, but that broke down, and while I was repairing it they used the little hand-pump. At the time she knocked against this berg, and reeled over, she leaked worse than she did before.

Question. While you laid up in Polaris Bay did the ship take any dam-age there from the ice that would have been likely to have made her leak ?

Answer. While she was lying on that berg her stem gave way. She cracked and there was a very bad place made in the bow, which we found out on the 23d of June. We could not find it out before on account of the water coming in and freezing over. When the ice melted away in the spring, we found out that she was leaking. Before, we did not know that she had received any damage. When the leak was dis-covered I was asked if I could take off the plate at her stem by the water-line six-foot mark, as she then was so much out of the water. I said " Yes," and took it off, and found one of the seams of the plank had given way. We put a lead patch over the cracked portion, and then I put the plate on.

Question. Was the stem cracked ?

Answer. Yes. On the port side there was a place that we could not get at on account of the water. That we could not fix and never did fix. I drove in some oakum, but could not get enough in to stop it altogether.

Question. She had been leaking then while you were drifting down the ice ?

Answer. Yes ; but she leaked much worse after we got that nip.

Question. Did you go out on the ice that evening when you were pass-ing things out ?

Answer. No. I was ordered to the pump and there I remained. Then I was ordered the second time ; I remained there until I was ordered to go down and get up steam. On the 15th we broke adrift, leaving a good many of our people on the ice. Of course the vessel's breaking adrift was purely accidental.

Question. At night what did you do ?

Answer. I was getting up steam, and pumping out the ship. That is what I was doing at the time. I was not on deck then. I could not get up there. We got up steam to pump, and the next morning I was on deck at 5 o'clock, and saw where we were—at Lifeboat Cove, above Littleton's Island. We were near the land on the east coast. The weather was then very clear; the wind had ceased, and the water was smooth. It was in a kind of bay—a bight in the land. We worked the vessel in with steam and sails, pumping all the while to keep the water down. We did this until we got her on the beach. We did not have a chance to see where or what the leak was when we got her on the beach.

We were satisfied that we could not put to sea in her again. If we had had plenty of coal we might have worked her out until we got down to Disco, Holsteinberg, or somewhere, where we could beach her higher than she was. When we went ashore we made up our minds it would be necessary to abandon her. We made up our minds that that was the last of her, and we would have to escape by other means. We then went to work to build a house, and lived there all winter. The next spring we built the boats. Mr. Chester, Mr. Coffin, and myself built the boats out of the roof of the cabin. The bottoms of the boats we made from the bunk-boards of the ship; all the rest was built from the roof of the cabin. They were flat-bottom boats. When the ice opened, on the 3d of June, we left Lifeboat Cove in these two boats, and started down to Sorfalik, where we made our first landing. We tried to get past a cape, but could not for the ice. We then had to put back to Sorfalik, a place that the natives call Etah Watana. We were in the boats about three weeks when we were picked up by the Ravenscraig. The Ravenscraig people were all very kind to us, and very glad to see us. From the Ravenscraig I was put on board the Intrepid on the 17th of July. I staid on board the Intrepid until the 23d of September. I was then transferred to the Eric, and by the Eric we were taken to Dundee. We reached Dundee on the 15th or 16th day of October. We remained there three days, and then went to Glasgow, and then sailed from Glasgow to New York, where we arrived November 6th, and I reported to the Brooklyn navy-yard, all the while that I was in Brooklyn.

Question. How were the engines and boilers when you beached the ship?

Answer. In first-class order. The engine was working better when we beached her than she ever was before.

Question. So that the steam department was in good order when you were compelled to run ashore?

Answer. Yes, sir.

Question. What did you think of the Polaris as a strong ship?

Answer. She was a very strong ship, and fit in every respect for the purpose of an Arctic voyage. She stood some pretty hard thumping, and if she had not been as strong a ship as she was, she would have gone down on the 4th of September.

Question. You never suffered for want of food on board the Polaris?

Answer. No, sir; nor anything else. The ship was well provided for an Arctic voyage.

Question. How long have you been in this country?

Answer. I have been eleven years in this country. I came here before I was of age. My home is in Brooklyn, New York. I am an American citizen.

Question. After Captain Hall's death, did you hear anybody say that he was glad he had died?

Answer. No, sir; every one I spoke to, always spoke very highly of Captain Hall, and were very sorry anything had happened to him.

Question. Did you use anything else to make steam with, but coal, while you were gone?

Answer. Only that night when we lost the men. We were compelled to get up steam in a hurry, and we had to use rosin and tar that we had on board; and wood and coal also.

Question. You had an apparatus on board to burn blubber with when you left New York?

Answer. Yes, sir; but we never used it. It was never used after it was used in Brooklyn. That was before I went on board of her; but I helped to put it together for them.